T0265412

Harmonic and Spectral Analysis

Harmonic and Spectral Analysis

$$f(x) = \int \widehat{f}(\chi)\chi(x)\, d\widehat{m}(\chi)$$

László Székelyhidi
University of Debrecen, Hungary

NEW JERSEY • LONDON • SINGAPORE • BEIJING • SHANGHAI • HONG KONG • TAIPEI • CHENNAI

Published by

World Scientific Publishing Co. Pte. Ltd.
5 Toh Tuck Link, Singapore 596224
USA office: 27 Warren Street, Suite 401-402, Hackensack, NJ 07601
UK office: 57 Shelton Street, Covent Garden, London WC2H 9HE

Library of Congress Cataloging-in-Publication Data
Székelyhidi, László, author.
Harmonic and spectral analysis / by László Székelyhidi (University of Debrecen, Hungary).
pages cm
Includes bibliographical references and index.
ISBN 978-9814531719 (hardcover : alk. paper)
1. Harmonic analysis. 2. Spectral sequences (Mathematics) I. Title.
QA403.S944 2014
515'.2433--dc23
2013048357

British Library Cataloguing-in-Publication Data
A catalogue record for this book is available from the British Library.

Copyright © 2014 by World Scientific Publishing Co. Pte. Ltd.

All rights reserved. This book, or parts thereof, may not be reproduced in any form or by any means, electronic or mechanical, including photocopying, recording or any information storage and retrieval system now known or to be invented, without written permission from the publisher.

For photocopying of material in this volume, please pay a copying fee through the Copyright Clearance Center, Inc., 222 Rosewood Drive, Danvers, MA 01923, USA. In this case permission to photocopy is not required from the publisher.

Printed in Singapore

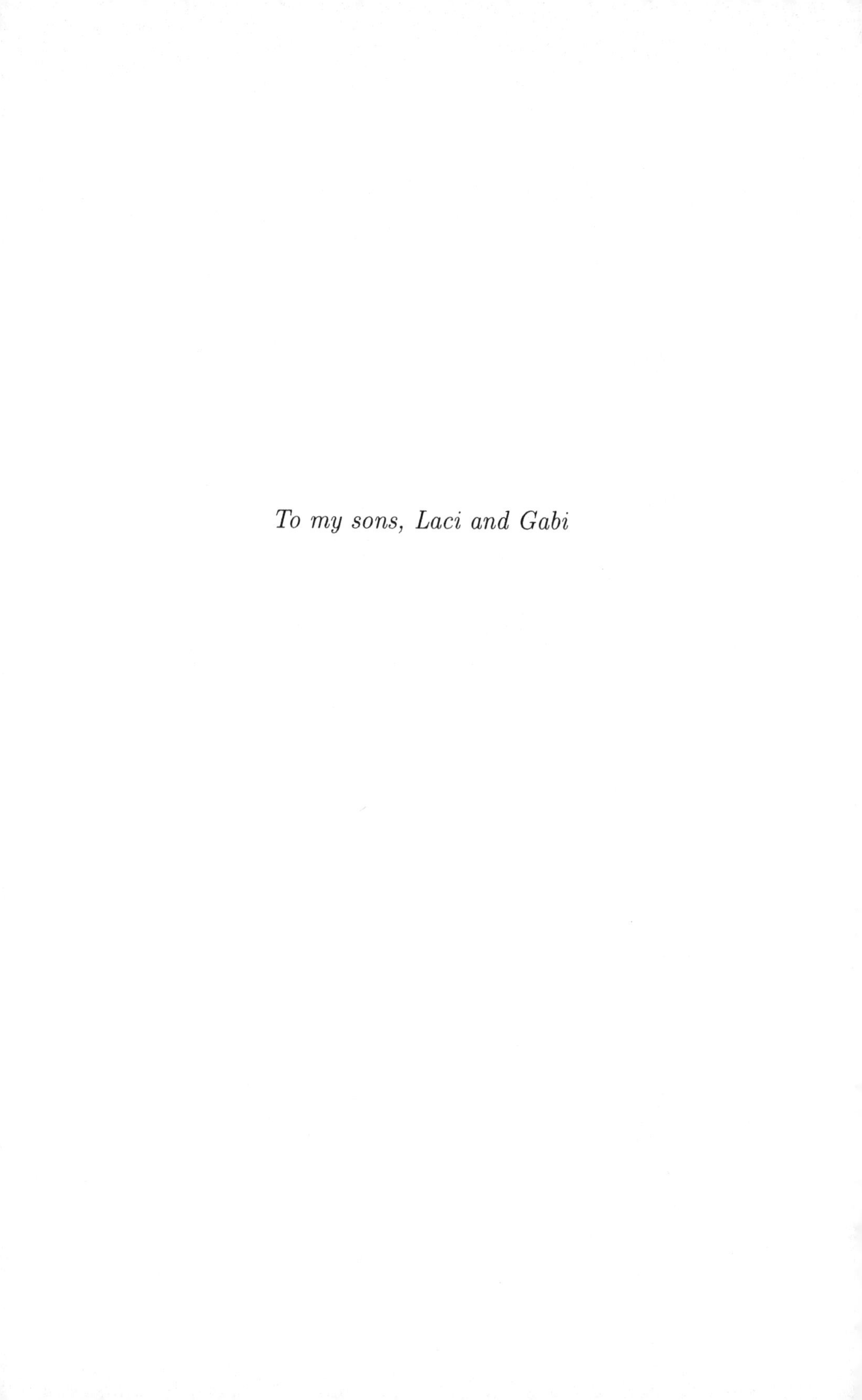

To my sons, Laci and Gabi

Contents

Preface

The roots of classical harmonic analysis go back to ancient times. The pioneer works of Fourier focused on the possibility of expanding functions into a series of basic harmonics. Nowadays researchers on abstract harmonic analysis have been studying functions, measures and spaces of such objects on topological groups. The purpose of this study is to build up these objects from elementary components, from special building bricks. These components serve as typical elements of the spaces in question, a kind of basis. Harmonic analysis means to discover the structure of the given space by finding those building bricks, and harmonic synthesis is the reconstruction process: describe and build up the space from the building bricks. A classical example is the following: the Uniqueness Theorem on the Fourier series of 2π-periodic continuous complex valued functions on the real line says that if the function is nonzero, then it has at least one nonzero Fourier coefficient. This can be reformulated by saying that the smallest translation invariant linear space including the function, and closed under uniform convergence, that is, the variety of the function, contains at least one complex exponential function. From Fejér's famous result on the uniform convergence of Fejér means to the function one derives the stronger property: the complex exponentials in the variety of the function actually span a dense subspace. The first reformulation above is a typical *harmonic analysis* result, while the second one is a *spectral synthesis* theorem. In the abstract setting all things become more general: instead of 2π-periodic continuous complex valued functions on the real line one considers complex valued functions defined on topological groups, and the role of complex exponentials is played by some special functions from a wider class. The tune, however, remains the same: are there any special functions available, and are there sufficiently many of them?

This book is divided into two parts. In the first part we present abstract harmonic analysis on locally compact Abelian groups. Our treatment is based on the duality theory, and on the structure theorems of such groups. Starting with the simplest case of finite Abelian groups we build up duality theory and harmonic analysis step by step on discrete, compact, elementary, compactly generated, and, finally, on arbitrary locally compact Abelian groups. The common feature of this theory is that one imposes different growth conditions on the functions in question: boundedness, square integrability, or integrability. Another specialty is that the building blocks of synthesis are the characters of the group, which underlines the importance of the group of all characters, that is, of the dual group. The basic questions of analysis and synthesis are reduced to the fundamental problems of duality theory: are there nontrivial characters at all, and are there sufficiently many characters on any locally compact Abelian group? These questions are answered in the positive by Pontryagin's Duality Theorem, which opens the door to the reconstruction of functions from their elementary components using Fourier transformation: on the top we are granted the beautiful Inversion Formula, the proper synthesis theorem together with Plancherel's Theorem, expressing the real strength and excellence of duality.

The second part of the book is devoted to spectral analysis and spectral synthesis. In some sense this is a generalization of classical harmonic analysis. Namely, one considers arbitrary continuous complex valued functions on locally compact Abelian groups without any growth conditions. Of course, the major part of the technical arsenal depending on duality and Fourier transformation is lost. Even the class of elementary components is not rich enough: easy examples show that the family of characters, that is, the dual group is not satisfactory to describe the variety of a given continuous function. In order to extend this family the idea is that one includes all "generalized" characters, which are the common eigenfunctions of all translation operators, and on the other hand, one takes into account the multiplicities of these eigenfunctions by introducing exponential monomials as basic building blocks. From now on the spectral analysis question reads as follows: is there any nonzero exponential monomial in the variety of a given function? And the corresponding synthesis problem is about the size of the set of exponential monomials in a given variety. The famous pioneering result in this area is due to Schwartz. His wonderful theorem states that each continuous complex valued function on the real line is the uniform limit on compact sets of exponential monomials taken from its variety. To prove this theorem Schwartz used deep results and strong tools from complex function theory, and obviously, this machinery is not available on topological groups. It turns out, however, that at least on discrete Abelian groups a whole lot of results are at our command from the theory of commutative rings. Classical theorems on Artin and Noether rings can be combined and incorporated in the theory resulting a method depending on the annihilators of varieties. This algebraic approach has been utilized on discrete Abelian groups and culminates in the characterization of those possessing spectral analysis and spectral synthesis.

Chapters 1 and 2 are about harmonic analysis on finite Abelian groups. These introductory chapters are to exhibit the basic ideas and problems of abstract harmonic analysis in a very simple situation, where some basic knowledge in linear algebra is sufficient to follow the lines. However, the reader is informed about the fundamental problems of duality, Fourier transformation and convolution without going into topological difficulties.

Chapter 3 is a survey on set theory and topology. Here we summarize the necessary requisites from these fields. Abstract set theory is not the real content of this book, however, we include here the most important concepts on partially ordered sets and generalized sequences to formulate the magic tool: Zorn's Lemma. In the topological introduction we start with general topological spaces and exhibit the most important properties of compact, connected and locally compact spaces, respectively. Here we also included Tikhonov's Theorem, Urishon's Lemma, and Tietze's Extension Theorem. We collected the basic information about topological groups, in general. In the final sections about topological vector spaces we study locally convex spaces, in particular, conjugate spaces and different versions of the Hahn–Banach Theorem. The final section is devoted to the Stone–Weierstrass Theorem.

In Chapter 4 a useful tool, the invariant mean, is introduced. Using the Banach–Alaoglu Theorem and the Markov–Kakutani Fixed Point Theorem we prove the existence of invariant means on discrete Abelian groups.

The first important step toward Pontryagin's Duality is taken in Chapter 5: we prove duality for discrete and for compact Abelian groups. In some sense, it is the heart of the first part of the book. In the following Chapter 6 we apply the previous results to extend duality for elementary Abelian groups.

Chapter 7 is the starting point of harmonic analysis on compact Abelian groups. The Riesz Representation Theorem plays an important role here, which makes it possible to introduce one of the most important tools of harmonic analysis: the Haar measure.

In Chapter 8 we are ready to prove Pontryagin's Duality Theorem in its full generality on locally compact Abelian groups. The proof depends on the famous Approximation Theorem of Pontryagin.

For harmonic analysis on arbitrary locally compact Abelian groups we need to obtain Haar measure on these groups. This is the content of Chapter 9. The existence proof depends again on the Markov–Kakutani Fixed Point Theorem. Using Haar measure and Haar integral we introduce convolution.

Harvest is coming in Chapter 10, the concluding chapter of the first part of the book. Using basic results from the theory of commutative algebras, in particular, from the Gelfand theory, we formulate the fundamental theorems of abstract harmonic analysis on locally compact Abelian groups. A detailed study of commutative algebras would exceed the scope of this volume, hence in these sections we have to omit most of the proofs. However, exhibiting the results may serve as a kind of motivation for the second part of our book.

Chapter 11 contains the introductory sections to the second part about spectral analysis and spectral synthesis. Here we collect the necessary definitions and tools from ring theory. The concepts of spectral analysis and spectral synthesis on varieties are defined, in connection with group representations and actions, as well. The basics of the annihilator technique are also settled.

Chapter 12 is the central part of our work. The results in these sections are relatively new, they have been published recently. These results may serve as the basis of further developments of spectral analysis and synthesis on groups, not necessarily only on commutative ones. Some of the results presented in this chapter have also been generalized to hypergroups by the author. Our main purpose here is to convert the problems of spectral theory from the varieties to the group algebra, making possible the extensive use of classical results in commutative ring theory. On the other hand, the characterization theorems for different function classes may present useful tools in the theory of functional equations.

Chapter 13 is devoted to the detailed study of torsion properties, in particular, to the torsion free rank of Abelian groups. It turns out that this concept plays a deciding role in spectral analysis and synthesis on discrete Abelian groups. On the other hand, it also has a close connection to the existence of "pathological" functions on the group.

In the two concluding chapters we present the most recent results on spectral analysis and synthesis for varieties on discrete Abelian groups. The two main characterization theorems on spectral analysis and on spectral synthesis, respectively, can be found in Chapter 14 and Chapter 15, respectively.

In this volume we have used some standard reference books, first of all [Hewitt and Ross (1979)] and [Rudin (1991)]. Most of the material in functional analysis in Chapter 3 has been taken from [Rudin (1991)]. For the topological background we have used [Kelley (1975)].

The first part of this book has been written during the author's stay at the University of Debrecen, while he was giving seminars on abstract harmonic analysis. We express our special thanks to the colleagues in Debrecen, who attended these seminars, and to the Department of Analysis for supporting our work.

The second part of the present volume has been completed at the University of Botswana, Gaborone, Botswana during the author's stay in the Spring, 2013, as a visiting professor. The author is indebted to the Department of Mathematics for providing a possibility of giving a seminar on the subject, and for the ideal circumstances.

J. Erdős

L. Fejér

And now let me express a personal comment. The setup of the first part of this book, in particular, the treatment of abstract harmonic analysis is based on the seminar talks delivered by my master, Jenő Erdős at the University of Debrecen, Hungary, in the late 70's. At that time I attended his seminars as a freshman, and he was the one who opened my mind and guided me through the wonderful world of harmonic analysis. I am deeply indebted to him and I dedicate this volume to his memory.

L. Székelyhidi

PART 1

Abstract Harmonic Analysis

Chapter 1

DUALITY OF FINITE ABELIAN GROUPS

1.1 Characters

In this chapter we introduce the basics of harmonic analysis on finite Abelian groups. This introduction serves as a presentation of the most important concepts, tools, ideas and results of abstract harmonic analysis in the simplest situation. Basic prerequisites are elementary algebra and linear algebra. For more references see e.g. [Spindler (1994a,b); Luong (2009); Isaacs (2006); Terras (1999); Halmos (1995, 1974a); van der Waerden (1991a,b)].

Let G be a finite Abelian[1] group written additively, with identity o. Let $|G|$ denote the number of elements of G, and let $\mathcal{C}(G)$ be the set of all complex valued functions on G. Clearly, $\mathcal{C}(G)$ is a complex Hilbert[2] space with pointwise addition and multiplication by scalars, further the inner product $\langle\, ,\rangle$ is defined by

$$\langle f, g\rangle = \frac{1}{|G|}\sum_{x\in G} f(x)\overline{g(x)}$$

for f, g in $\mathcal{C}(G)$. We may denote this space by $L^2(G)$, which is obviously isometrically isomorphic to the Hilbert space $\mathbb{C}^n$, and the isomorphism is provided by the mapping $f \mapsto [f(x_1), f(x_2), \ldots, f(x_n)]$, where $x_1, x_2, \ldots, x_n$ are the elements of G. An orthogonal basis of $L^2(G)$ is formed by the characteristic functions of singletons. The characteristic function of an arbitrary set A will be denoted by δ_A, which takes the value 1 at the points of A, and 0 otherwise. If $A = \{a\}$ is a singleton, then we simply write δ_a for δ_A.

[1]Niels Henrik Abel, Norwegian mathematician (1802-1829)

[2]David Hilbert, German mathematician (1862-1943)

N. H. Abel

D. Hilbert

Homomorphisms of G into the multiplicative group $\mathbb{T}$ of complex numbers of modulus 1 are called *characters*.

Theorem 1.1. *The characters of a finite Abelian group are homomorphisms of the group into the multiplicative group of complex n-th roots of unity, where n is the number of elements of the group.*

Proof. For each x in G we have $nx = o$, hence for every character χ it follows $\chi(x)^n = \chi(nx) = \chi(o) = 1$. □

Theorem 1.2. *Different characters of a finite Abelian group are orthogonal, and their norm is* 1.

Proof. First we show that if χ is a character of the finite Abelian group G, then

$$\sum_{x\in G} \chi(x) = \begin{cases} 0 & \text{if}\, \chi \neq 1, \\ |G| & \text{if}\, \chi = 1\,. \end{cases}$$

Indeed, we have for each y in G

$$\sum_{x\in G} \chi(x) = \sum_{x\in G} \chi(x+y) = \sum_{x\in G} \chi(x)\chi(y) = \chi(y) \sum_{x\in G} \chi(x)\,,$$

which gives the statement.

Now let χ_1, χ_2 be different characters of G, then $\chi = \chi_1 \cdot \overline{\chi}_2$ is a character different from 1, which implies orthogonality, by the previous observation. The statement about the norm is obvious. □

We define the *translation operator* τ_y corresponding to the element y in G by $\tau_y f(x) = f(x-y)$ for every function f in $\mathcal{C}(G)$ and each element x in G. The function $\tau_y f$ is called the *translate* of f by y.

We call the function f in $\mathcal{C}(G)$ *normed*, if $f(o) = 1$. Obviously, this is, in general, different from the concept of a function of norm 1 in $L^2(G)$.

Theorem 1.3. *Translation operators of a finite Abelian group G are commuting unitary operators of the Hilbert space $L^2(G)$.*

Proof. As G is Abelian, translation operators obviously commute, and τ_{-y} is the inverse of τ_y. Further we have

$$\langle \tau_y f, g \rangle = \frac{1}{|G|} \sum_{x \in G} f(x-y)\overline{g(x)} = \frac{1}{|G|} \sum_{x \in G} f(x)\overline{g(x+y)} = \langle f, \tau_{-y} g \rangle$$

for each y in G and f, g in $\mathcal{C}(G)$, which shows that the adjoint of τ_y is equal to its inverse, that is, τ_y is unitary. □

Theorem 1.4. *The common normed eigenfunctions of all translation operators of a finite Abelian group are precisely the characters.*

Proof. Let f be a common normed eigenfunction of all translation operators. Then for each y in G there is a complex number $\lambda(y)$ such that

$$f(x+y) = \lambda(y) \cdot f(x)$$

holds for each x in G. As $f(o) = 1$, substitution $x = o$ gives $f = \lambda$, thus f is a character. Conversely, it is clear that every character is a common normed eigenfunction of all translation operators. □

We see from this proof that the character χ is the eigenfunction of the translation τ_y belonging to the eigenvalue $\overline{\chi(y)}$.

Theorem 1.5. *All characters of a finite Abelian group G form an orthonormal basis of the Hilbert space $L^2(G)$.*

Proof. It is known from linear algebra (see e.g. [Spindler (1994a)], Proposition (9.28), p. 187) that given a set of commuting unitary operators in a finite dimensional Hilbert space, then there exists an orthonormal basis consisting of common eigenvectors of these operators. In our case the common eigenfunctions are exactly the characters. By orthogonality it follows that every character occurs in this basis. □

By the theorem, the number of all characters is equal to the dimension of $L^2(G)$, that is, to $|G| = n$.

1.2 Dual group

As it is easy to see, all characters form an Abelian group with respect to pointwise multiplication, which is called the *dual group*, or simply the *dual* of G. We denote it by $\widehat{G}$. We have seen that $|\widehat{G}| = |G|$. In the dual group the inverse of the character χ is $\overline{\chi}$, the complex conjugate of χ, which is actually the reciprocal of the function χ.

Theorem 1.6. *The dual of a finite cyclic group is isomorphic to the group itself.*

Proof. Let x be the generator of the cyclic group of order n, and let χ be an arbitrary character. Then we obviously have $\chi(kx) = \chi(x)^k = \alpha^k$ for $k = 1, 2, \ldots, n$, where α is a complex number of modulus 1, moreover, by

$$\alpha^n = \chi(x)^n = \chi(nx) = \chi(o) = 1\,,$$

α is an n-th root of unity. Clearly, the mapping $\chi \mapsto \alpha$ is an isomorphism of the dual group onto the multiplicative group of n-th roots of unity. On the other hand, it is obvious that the cyclic group of order n is also isomorphic to the multiplicative group of n-th roots of unity. □

Although every finite cyclic Abelian group is isomorphic to its dual, but it is clear from the above proof that the isomorphism is far from being "natural", as it does not reflect anything about the relation between the group and its dual. It is just a trivial consequence of the simple fact that any two cyclic groups of the same order are isomorphic.

The dual of the finite Abelian group G has also its dual: $\widehat{\widehat{G}}$, the *second dual* of G. There is a natural way to set up a mapping of G into its second dual: for each x in G let Φ_x be the character of $\widehat{G}$ defined by $\Phi_x(\chi) = \chi(x)$ for each χ in $\widehat{G}$. The mapping $x \mapsto \Phi_x$ is called the *canonical homomorphism*, or *natural homomorphism* of G into its second dual. The most important result in the duality theory of finite Abelian groups is the following.

Theorem 1.7. *Finite Abelian group is isomorphic to its second dual via the canonical homomorphism.*

Proof. Obviously, the canonical homomorphism is a homomorphism, indeed. To prove injectivity one considers the following *character table*

$$\begin{pmatrix} \chi_1(x_1) & \chi_1(x_2) & \dots & \chi_1(x_n) \\ \chi_2(x_1) & \dots & \dots & \dots \\ \dots & \dots & \dots & \dots \\ \chi_n(x_1) & \chi_n(x_2) & \dots & \chi_n(x_n) \end{pmatrix}$$

where $\chi_1, \chi_2, \dots, \chi_n$ are all the different characters of G. If all characters take the value 1 at some element $x \neq o$, then the columns in this matrix corresponding to x and to o are identical, hence the matrix is singular, contradicting the linear independence of the characters, which is an immediate consequence of their orthogonality. It follows that the kernel of the canonical homomorphism is the identity, which implies its injectivity. Surjectivity is a direct consequence of the equal cardinalities of the two groups. □

Theorem 1.7 is called Pontryagin's[3] Duality Theorem for finite Abelian groups. The property of G that for each element different from the identity there is a character, whose value at this element is different from 1 is expressed by saying that the group has *sufficiently many* characters.

Clearly, this property can be formulated in the way that for any two different elements there exists a character, which takes different values at the two elements. We can say shortly that the characters *separate* the elements of the group, or, the characters form a *separating family* for the group. The above argument shows that this property is necessary and sufficient for the canonical homomorphism is injective.

L. S. Pontryagin

[3]Lev Semenovich Pontryagin, Russian mathematician (1908-1988)

Chapter 2

HARMONIC ANALYSIS ON FINITE ABELIAN GROUPS

2.1 Fourier transformation

Fourier[1] transformation is a basic tool of modern analysis. In this section we introduce it on finite Abelian groups and exhibit its most important properties.

Let G be a finite Abelian group. In the basis formed by the characters each function f in $L^2(G)$ has a unique representation of the form

$$f = \sum_{k=1}^{n} \lambda_k \chi_k \,,$$

and, by the orthonormality of the characters, the coefficients λ_k can be computed by $\lambda_k = \langle f, \chi_k \rangle$. These are called the *Fourier coefficients* of the function f, which obviously satisfy

$$f = \sum_{\chi \in \widehat{G}} \langle f, \chi \rangle \chi \,.$$

J-B. J. Fourier

To each function in $L^2(G)$ we assign the *sequence of Fourier coefficients* occuring in the representation above: for each function f in $L^2(G)$ and character χ in $\widehat{G}$ we let $\widehat{f}(\chi) = \langle f, \chi \rangle$. The complex valued function $\widehat{f}$ defined on the dual of G is the *Fourier transform* of f, and the mapping $f \mapsto \widehat{f}$ is called *Fourier transformation*. By the above

[1] Jean-Baptiste Joseph Fourier, French mathematician (1768-1830)

consideration we have

$$\widehat{f}(\chi) = \frac{1}{|G|} \sum_{x \in G} f(x)\overline{\chi(x)}\,,$$

whenever f is a complex valued function, and χ is a character on G.

For our later purposes we introduce an inner product on the set of all functions defined on the dual of G as follows:

$$\langle \varphi, \psi \rangle_{\wedge} = \sum_{\chi \in \widehat{G}} \varphi(\chi)\overline{\psi(\chi)}\,,$$

where φ, ψ are arbitrary functions on the dual of G.

The following theorem, due to Plancherel[2], expresses one of the most important properties of the Fourier transformation.

Theorem 2.1. *(Plancherel's Theorem) The Fourier transformation on the finite Abelian group G is an isometric isomorphism of $L^2(G)$ onto $L^2(\widehat{G})$.*

Proof. It is easy to check that the Fourier transformation is a homomorphic mapping. On the other hand,

$$\langle \widehat{f}, \widehat{g} \rangle_{\wedge} = \sum_{\chi \in \widehat{G}} \widehat{f}(\chi)\overline{\widehat{g}(\chi)}$$

$$= \sum_{\chi \in \widehat{G}} \left[\frac{1}{|G|} \sum_{x \in G} f(x)\overline{\chi(x)} \frac{1}{|G|} \sum_{y \in G} \overline{g(y)}\chi(y) \right]$$

$$= \frac{1}{|G|^2} \sum_{\chi \in \widehat{G}} \left[\sum_{x \in G} \sum_{y \in G} f(x)\overline{g(y)} \right] \overline{\chi(x-y)}$$

$$= \frac{1}{|G|^2} \sum_{\chi \in \widehat{G}} \sum_{x \in G} f(x) \left[\sum_{y \in G} \overline{g(y)} \sum_{\chi \in \widehat{G}} \overline{\chi(x-y)} \right]$$

$$= \frac{1}{|G|} \sum_{x \in G} f(x)\overline{g(x)} = \langle f, g \rangle$$

holds for each f, g in $L^2(G)$. Here we have used the fact that $\sum_{\chi \in \widehat{G}} \chi(x) = |G|$, for $x = o$ and 0 otherwise, which comes from the first part of the proof of Theorem 1.2, by the observation that $\chi \mapsto \chi(x)$ is a character of $\widehat{G}$ for each x in G. This implies that the Fourier transformation is an isometry, hence it is injective. Surjectivity follows from the equality of the dimensions of $L^2(G)$ and $L^2(\widehat{G})$: in both cases it is $|G|$. □

[2]Michel Plancherel, Swiss mathematician (1885-1967)

An important consequence of Plancherel's Theorem is Parseval's Formula[3]:

$$\langle f, f \rangle = \langle \widehat{f}, \widehat{f} \rangle_{\wedge}, \tag{2.1}$$

which holds for each f in $L^2(G)$, expressing the isometric property of the Fourier transformation. The inverse mapping of the Fourier transformation is an isometric isomorphism, too, by Plancherel's Theorem. A natural question is whether the inverse transform can be described by an explicit formula. A fundamental property of the Fourier transformation is that it can be expressed by the so-called *Inversion Formula*, which is similar to the one defining the Fourier transform.

M. Plancherel

Theorem 2.2. *(Inversion Theorem) The inverse of the Fourier transformation on the finite Abelian group G is given by*

$$f(x) = \sum_{\chi \in \widehat{G}} \widehat{f}(\chi)\chi(x),$$

whenever f is in $L^2(G)$, and x is in G.

Proof. This formula is exactly the expansion of the function f in the orthonormal basis formed by all characters. □

2.2 Convolution

In addition to pointwise linear operations also pointwise multiplication is defined in the set of Fourier transforms. Now we try to figure out which is the corresponding operation in $L^2(G)$. For this purpose we introduce a new concept.

Given a function f we consider the formal linear combination $\sum_{x \in G} f(x)x$. As it is easy to see the set of all such combinations forms an algebra, the

[3]Marc-Antoine Parseval des Chênes, French mathematician (1755-1836)

so-called *group algebra*, if we define the operations as follows:

$$\sum_{x\in G} f(x)x + \sum_{y\in G} g(y)y = \sum_{z\in G}[f(z)+g(z)]z\,,$$

$$\lambda\cdot\sum_{x\in G} f(x)x = \sum_{x\in G}\lambda f(x)x\,,$$

and

$$\sum_{x\in G} f(x)x\cdot\sum_{y\in G} g(y)y = \sum_{x\in G}\sum_{y\in G} f(x)g(y)xy\,.$$

Multiplication can also be written in the following way:

$$\sum_{x\in G} f(x)x\cdot\sum_{y\in G} g(y)y = \sum_{z\in G}\sum_{xy=z} f(x)g(y)z$$

$$= \sum_{x\in G}\Big[\sum_{y\in G} f(xy^{-1})g(y)\Big]x\,,$$

which means that the product of the formal linear combinations corresponding to f and g in this algebra is the formal linear combination corresponding to the function defined by the formula $x\mapsto\sum_{y\in G} f(xy^{-1})g(y)$. The *convolution* of f and g at x is defined as this value multiplied by the factor $\frac{1}{|G|}$. In other words, we define

$$f*g(x) = \frac{1}{|G|}\sum_{y\in G} f(xy^{-1})g(y)\,.$$

Obviously, $f*g=g*f$. The following theorem is of fundamental importance.

Theorem 2.3. *(Convolution Theorem) For arbitrary complex valued functions f,g on a finite Abelian group we have*

$$(f*g)\widehat{\ } = \widehat{f}\cdot\widehat{g}\,.$$

Proof. Let x be arbitrary in G, then it follows

$$(f*g)\widehat{\ }(\chi) = \langle f*g,\chi\rangle = \frac{1}{|G|}\sum_{x\in G}(f*g)(x)\overline{\chi(x)}$$

$$= \frac{1}{|G|^2}\sum_{x\in G}\sum_{y\in G} f(x-y)g(y)\overline{\chi(x)}$$

$$= \frac{1}{|G|^2}\sum_{x\in G}\sum_{y\in G} f(x-y)\overline{\chi(x-y)}g(y)\overline{\chi(y)}$$

$$= \frac{1}{|G|^2}\sum_{x\in G}\sum_{y\in G} f(z)\overline{\chi(z)}g(y)\overline{\chi(y)}$$

$$= \frac{1}{|G|}\sum_{z\in G} f(z)\overline{\chi(z)}\frac{1}{|G|}\sum_{y\in G} g(y)\overline{\chi(y)} = \widehat{f}(\chi)\widehat{g}(\chi)$$

for each character χ. □

2.3 Convolution operators

In the usual way, given a Hilbert space H we denote by $\mathcal{L}(H)$ the set of all linear operators on H. It is well-known that $\mathcal{L}(H)$, equipped with the usual operator norm is a Banach[4] space. For our purposes the most important case is the one, where H is the $L^2(G)$ Hilbert space.

Now we introduce convolution operators. For each a in $\mathcal{C}(G)$ the mapping defined by

$$A_a : f \mapsto a * f$$

is called the *convolution operator* corresponding to the function a. Let a^* denote the function defined by

$$a^*(x) = \overline{a(-x)}$$

whenever x is in G.

S. Banach

Theorem 2.4. *Let G be a finite Abelian group. For each function a in $\mathcal{C}(G)$ the corresponding convolution operator A_a is a linear operator of the Hilbert space $L^2(G)$ with adjoint operator A_{a^*}. In particular, A_a is a self-adjoint operator if and only if $a^* = a$. All convolution operators form a commuting family.*

Proof. The linearity of convolution operators is obvious. For the adjoint of A_a we have, by definition,

$$\langle A_a f, g\rangle = \langle f, A_a^* g\rangle$$

for each f, g in $L^2(G)$. Let u be arbitrary in G and $g = \delta_u$, then by the previous equation we infer

$$a * f(u) = \sum_{x\in G} f(x)\overline{A_a^*\delta_u(x)}$$

whenever f is in $L^2(G)$ and u is in G. As this holds for each f it follows

$$A_a^*\delta_u(x) = \frac{1}{|G|}a(u-x)$$

for each x, u in G. As every function g in $L^2(G)$ can be written in the form

$$g = \sum_{u\in G} g(u)\delta_u\,,$$

[4]Stefan Banach, Polish mathematician (1892-1945)

hence, by the previous equation, and, by the linearity of A_a^*, we have

$$A_a^* g(x) = \sum_{u\in G} g(u) A_a^* \delta_u(x) = \frac{1}{|G|} \sum_{u\in G} a(u-x) g(u)$$

$$= \frac{1}{|G|} \sum_{u\in G} a^*(x-u) g(u) = a^* * g(x) \,,$$

and this is our statement for the adjoint operator.

The commuting property of convolution operators is an immediate consequence of the commutativity of the convolution. □

By the commuting property of convolution operators, and, by the statement proved for their adjoint operators, it follows that the convolution operators are normal operators of the Hilbert space $L^2(G)$.

We note that all translation operators are convolution operators, as for each y in G and f in $L^2(G)$ we have

$$\tau_y f(x) = f(x-y) = \sum_{u\in G} f(x-u)\delta_y(u)$$

$$= \sum_{u\in G} \delta_y(x-u) f(u) = |G|\, \delta_y * f(x) \,,$$

hence

$$\tau_y = A_{|G|\delta_y}, \quad \text{or} \quad A_{\delta_y} = \frac{1}{|G|}\tau_y \,.$$

Moreover, as we have seen before, these operators are even unitary operators.

From the above considerations it is clear that the commuting character of convolution operators, which plays a fundamental role in harmonic analysis, is a consequence of the commutativity of the group. This is obvious for the translation operators, however, every convolution operator is a linear combination of translation operators. Indeed, for every a in $\mathcal{C}(G)$ it follows

$$a = \sum_{y\in G} a(y)\delta_y \,,$$

hence we have

$$A_a = \sum_{y\in G} a(y) A_{\delta_y} = \frac{1}{|G|} \sum_{y\in G} a(y)\tau_y \,.$$

This also implies that every linear operator, commuting with all translation operators, commutes with all convolution operators, too. The following theorem shows that, in fact, this property characterizes convolution operators on $L^2(G)$.

Theorem 2.5. *Let G be a finite Abelian group. A linear operator on $L^2(G)$ is commuting with all translation operators if and only if it is a convolution operator.*

Proof. The sufficiency of the condition has been verified above. Suppose now that B is a linear operator on the Hilbert space $L^2(G)$ such that
$$\tau_y B = B\tau_y$$
holds for each y in G. This means that
$$(\tau_y B)\delta_o = B(\tau_y\delta_o) = B\delta_y$$
holds whenever y is in G. On the other hand, we have seen above that
$$(\tau_y B)\delta_o = \tau_y(B\delta_o) = |G|(\delta_y * B\delta_o) = |G|B\delta_o * \delta_y\,,$$
hence
$$B\delta_y = |G|B\delta_o * \delta_y$$
holds for each y in G. By the linearity of B and that of the convolution, we have the obvious equation
$$f = \sum_{y\in G} f(y)\delta_y\,,$$
which holds for each f in $L^2(G)$. Finally, this implies that
$$Bf = |G|B\delta_o * f\,,$$
that is, $B = A_{|G|B\delta_o}$, a convolution operator. □

We have seen that the characters of the finite Abelian group G are exactly the common normed eigenfunctions of all translation operators. By the following result the characters are also the common eigenfunctions of all convolution operators.

Theorem 2.6. *Let G be a finite Abelian group. The common normed eigenfunctions of all convolution operators are exactly the characters of G.*

Proof. As the translation operators are convolution operators, it is enough to show that every character is an eigenfunction of each convolution operator. Let χ be a character and a a function in $\mathcal{C}(G)$, then we have
$$A_a\chi(x) = a * \chi(x) = \frac{1}{|G|}\sum_{u\in G} a(u)\chi(x-u) = \frac{1}{|G|}\sum_{u\in G} a(u)\chi(-u)\cdot\chi(x)$$
for each x in G, that is, χ is an eigenfunction of A_a. □

Chapter 3

SET THEORY AND TOPOLOGY

3.1 Basics from set theory

In this section we summarize the basic knowledge from set theory, which we shall use in the sequel without special reference. For the necessary information the reader should refer to [Hausdorff (1962); Kuratowski (1962); Jech (1973); Halmos (1974b)].

We assume that the reader is familiar with the basic operations used in the so-called naive set theory, like Boolean[1] operations, Descartes[2]–products and their fundamental properties. Although we treat set theory tacitly as an axiomatic discipline with the two basic concepts *"set"* and *"element of"*, we do not intend to go into the details of the axiomatic treatment. Here we want to focus only on some particular concepts and results, which are indispensable from the point of view of our investigations.

G. Boole

R. Descartes

[1]George Boole, English mathematician (1815-1864)
[2]René Descartes, French philosopher (1596-1650)

An *equivalence relation* R on the set X is a relation $R \subseteq X \times X$ with the following properties:

1. R is *reflexive*, that is, (i, i) belongs to R for each i in X;
2. R is *symmetric*, that is, if for some i, j in X (i, j) belongs to R, then (j, i) belongs to R, too;
3. R is *transitive*, that is, if for some i, j, k in X (i, j) and (j, k) belong to R, then (i, k) belongs to R, too.

For each x in X we denote by $R(x)$ the set of all elements y in X for which (x, y) belongs to R, which is called the *class* of x. It is easy to see that the classes of two elements are either disjoint, or they coincide, further the union of all classes is X. This property can be expressed by saying that the classes of all elements form a *partition* of X. The set of all classes is called the *factor set* of X with respect to R, and it is denoted by X/R. The mapping, which assigns to each element of X its class is called the *natural mapping* of X onto its factor set X/R.

The set I is called a *partially ordered* set, if a *partial ordering* R is defined on it, which is a relation $R \subseteq I \times I$ satisfying the following properties:

1. R is reflexive;
2. R is *antisymmetric*, that is, if for some i, j in I the pairs (i, j) and (j, i) belong to R, then $i = j$;
3. R is transitive.

For example, the *set-theoretical inclusion* $A \subseteq B$ is obviously a partial ordering on any family I of sets, but this fails to hold on the *membership* $A \in B$ of sets, as it, in general, does not have any of the above properties. In this work, when dealing with partially ordered sets the fact that (i, j) belongs to R will often be denoted by $i \leqslant j$.

The partially ordered set I is called a *chain*, if it has the following additional property: any two elements i, j of I are *comparable* in the sense of the relation R, that is, either of the two pairs (i, j) and (j, i) belongs to R. In this case we say that I is a *linearly ordered*, or simply *ordered* set, and the relation R is an *ordering*.

Any subset of a partially ordered set is obviously a partially ordered set, equipped with the restriction of the given partial ordering to the subset. The subsets of a partially ordered set will always be considered as partially ordered sets with this ordering.

Given a subset H in the partially ordered set I an element i in I such that $h \leqslant i$ holds for each h in H is called an *upper bound* of H.

By a *maximal element* in a partially ordered set I we mean an element m of I such that there is no element $i \neq m$ in I such that $m \leqslant i$. A partially ordered set obviously may have more than one maximal elements, and it may also happen that it has none at all. However, in any chain there is at most one maximal element. We note that the maximal elements of a subset of a partially ordered set always belong to the subset, but this is not necessarily true for its upper bounds.

Let I be a partially ordered set, in which for any pair i, j of elements there is a k in I with $i \leqslant k$ and $j \leqslant k$. Then I is called a *directed set*. Functions defined on directed sets are called *Moore*[3]*–Smith*[4] *sequences*, or *generalized sequences*. Sometimes they are also called *nets*. For instance, the set of natural numbers equipped with the natural ordering is obviously partially ordered and directed, and the generalized sequences on it are exactly the ordinary sequences. The generalized sequence x on the directed set I is sometimes denoted by $(x_i)_{i\in I}$, similarly to the ordinary sequences. Obviously, we also have to specify the range of the generalized sequence. If this range is a subset of the set H, that is $x : I \to H$ is a function, then we say that x is a generalized sequence *in the set* H. Generalized sequences play the same role in non-metric topological spaces as the ordinary sequences in metric spaces. Here we do not intend to go into the details of the theory of generalized sequences, just we need some simple concepts and properties concerning them.

The most famous, however most debated axiom of set theory is the *Axiom of Choice*. A possible formulation of this axiom follows.

The Axiom of Choice: *Let I be a nonempty set, and let $(A_i)_{i\in I}$ be a family of nonempty sets. Then there exists a function $f : I \to \bigcup_i A_i$ such that $f(i)$ belongs to A_i for each i in I.*

Concerning the Axiom of Choice the reader is referred to [Moore (1982)] and the references given there. One of the most important application of the Axiom of Choice is related to the product of a not necessarily finite family of sets. Let I be a set, and let $(A_i)_{i\in I}$ be a family of sets. We

---

[3]Eliakim Hastings Moore, American mathematician (1862-1932)
[4]Herman Lyle Smith, American mathematician (1892-1950)

denote by $A = \Pi_{i\in I} A_i$ the collection of all functions $f : I \to \bigcup_{i\in I}$ with the property that $f(i)$ belongs to A_i for each i in I. The set A is called the *product* of the sets $(A_i)_{i\in I}$. By the above formulation of the Axiom of Choice it is clear that we have the following result.

Theorem 3.1. *The product of a nonempty family of nonempty sets is nonempty.*

For each j in I the function $p_j : A \to A_j$ defined on the product A of the sets $(A_i)_{i\in I}$ by $p_j(f) = f(j)$, whenever f is in A, is called the *projection* corresponding to the j-th factor, or *j-th projection.*

The Axiom of Choice has a number of different equivalent formulations, which, of course, are theorems in our treatment. One of them, which will be used frequently in the sequel, is the following (for the original see [Zorn (1935)]).

Theorem 3.2. *(Zorn's[5] Lemma) If in a partially ordered set every chain has an upper bound, then the set has a maximal element.*

3.2 Topological background

In our investigations so far we have considered an Abelian group as the basic structure, which is a pure algebraic concept. However, the classical theory of harmonic analysis has grown up from the study of continuous periodic functions on the real line, hence we might expect that we have to open doors to analysis. In what follows we shall consider topological groups, and we need some basic knowledge from general topology. For basic references the reader should consult with [Bourbaki (1998a,b); Kelley (1975); Kuratowski (1962); Munkres (1975); Mendelson (1990)].

By a *topology* on a set X we mean a family of subsets of X, which contains the intersection of any two members of it, and the union of any subfamily of it, further it contains X and the empty set, too. The set X equipped with a topology is called a *topological space*, and the sets belonging to the family defining the topology are called *open sets*, while their complements are called *closed sets*. The *closure* Y^{cl} of a subset Y in the topological space X is the intersection of all closed sets containing Y, and its *interior* is the union of all open subsets of Y. As it is easy to see, in any

[5]Max August Zorn, German mathematician (1906-1993)

topological space the closure of a set is always closed, and the interior of a set is always open. Further, a subset is closed if and only if it is equal to its closure, and it is open if and only if it is equal to its interior. A subset is called *dense*, if its closure is the whole space.

A topological space is called *discrete*, if every subset is open.

As it is easy to see, the intersection of any nonempty set of topologies given on the set X is a topology on X. It follows that given any family of subsets of X it makes sense to consider the intersection of all topologies containing this family, and to call it the topology *generated* by the given family of sets. In this case the family itself is called a *sub-basis* of the generated topology. Hence, in order that a family of sets is a sub-basis of a topology it is necessary and sufficient that every nonempty open set is the union of finite intersections of members of the family. If every open set is the union of the members of a family of sets, then this family is called a *basis* of the topology.

A point of a topological space is called an *interior point* of a subset, if it belongs to the interior of the subset, and a point of the interior of the complement is called an *exterior point* of the subset. A point of the space is a *boundary point* of a subset, if it is neither an interior nor an exterior point of the subset. A point of a topological space is called a *limit point* of a subset, if every neighborhood of the point contains an element of the set, which is different from the given point. A point of a topological space is called an *isolated point* of a subset, if there is a neighborhood of the point, whose intersection with the set is the singleton consisting of the given point. A subset in a topological space is a *neighborhood* of a point, if the point is an interior point of the subset. A family of neighborhoods of a point is called a *neighborhood basis* of the point, if every neighborhood of the point includes a member of this family. Obviously, every point in a topological space has a neighborhood basis consisting of open neighborhoods. It is also clear that for each point x in a topological space every set in a neighborhood basis $\mathcal{U}(x)$ of x contains x, and the intersection of any two members of $\mathcal{U}(x)$ contains an element of $\mathcal{U}(x)$.

Given a subset Y in the topological space X the intersections of Y with the open sets of X make a topology of Y, and Y equipped with this topology is called a *subspace* of X.

Given two topologies on a set one is called *weaker* than the other, if every set, which is open in the first, is open in the second, too. In this case the second topology is called *stronger* than the first. Given a family of subsets of the set X there is a weakest topology on X in which the given sets are all open, and it is easy to see that it is made up by the unions of the finite intersections of the given subsets together with the empty set. This is the topology generated by the given family of sets – as it is easy to see.

We say that a mapping from a topological space X into another topological space Y is *continuous*, if the inverse image of each open set in Y is open in X. A mapping from X into Y is *open*, if the image of each open set in X is open in Y, and the mapping is *closed*, if the image of each close set in X is closed in Y. A bijective mapping of X onto Y is called a *homeomorphism*, if it is continuous, and its inverse is also continuous. We say that X and Y are *homeomorphic*, if there is a homeomorphism from X onto Y.

The mapping f of the topological space X into the topological space Y is said to be *continuous at the point* x_0 in X, if the inverse image of every neighborhood of the point $f(x_0)$ by the mapping f is a neighborhood of x_0. It is clear that the mapping f of the topological space X into the topological space Y is continuous if and only if it is continuous at every point of the space X. It is also easy to see that if the mapping f of the topological space X into the topological space Y is continuous at the point x_0 of X, and the mapping g of the topological space Y into the topological space Z is continuous at the point $f(x_0)$ of Y, then the composite mapping $g \circ f$ of the topological space X into the topological space Z is continuous at the point x_0.

If I is a set, and for each i in I the mapping f_i maps the set Y into the topological space X_i, then there is a weakest topology on Y such that if Y is equipped with this topology, then every f_i is continuous. As it is easy to see, this topology is generated by the inverse images of the open sets in X_i by the mappings. This topology is called the topology *induced by the system of functions* f_i.

The product set X of the topological spaces X_i equipped with the topology induced by the projections of X onto the sets X_i, is called the *product of the topological spaces* X_i. It is easy to see that if the sets $\mathcal{B}_i$ form a basis of the topology of X_i for each i in I, then the sets $\prod_{i\in I} B_i$ form a basis of the topology of the product space, where B_i is in $\mathcal{B}_i$ and $B_i = X_i$ for all but a finite set of the i's.

Given an equivalence relation R on the topological space X those subsets A of the factor set X/R, whose inverse images by the natural mapping are open in X give a topology on the factor set, which is called the *factor topology* of the factor set X/R. As it is easy to see, the natural mapping of X onto the factor space X/R is continuous.

We say that the family $(A_i)_{i\in I}$ of subsets of the set X is a *covering* of the subset A in X, if $A \subseteq \bigcup_{i\in I} A_i$. It is called a *finite covering*, if the family $(A_i)_{i\in I}$ is a finite set. If X is a topological space, then the covering $(A_i)_{i\in I}$ is called an *open covering*, if all the sets A_i are open.

We say that the nonempty family $(A_i)_{i\in I}$ of subsets of set X is *centered*, if the intersection of each nonempty subfamily of it is nonempty. Hence a nonempty family of subsets of a set X is centered if and only if each finite subfamily of the complements of the sets in this family is not a covering of X.

We say that the subset A of a topological space is *compact*, if every open covering of A contains a finite subcovering. We say that the topological space X is a *compact topological space*, if it is a compact subset of itself. As it is easy to see, a subset A of the topological space X is compact if and only if it is, as a subspace, a compact topological space. We have the following theorem.

Theorem 3.3. *A topological space is compact if and only if every centered family of its closed subsets has a nonempty intersection.*

Proof. Let $(A_i)_{i\in I}$ be a centered family of closed subsets in the compact topological space X, and suppose that $\bigcap_{i\in I} A_i = \varnothing$. Then the family $(A_i^c)_{i\in I}$ of the complements is an open covering of X, which has no finite subcovering, and this contradicts the compactness of X.

Conversely, suppose that the condition of the theorem is satisfied, and let $(U_i)_{i\in I}$ be an open covering of X. Then the closed complements $(U_i^c)_{i\in I}$ form a centered family having empty intersection, hence this family cannot be centered, that is, it has a finite subfamily with empty intersection. The complements of the sets in this finite subfamily form a finite subcovering of the original open covering. □

Theorem 3.4. *(Tikhonov's[6] Theorem) The product of compact topological spaces is compact.*

Proof. Let the centered family $\mathcal{C}$ of closed sets in the product X of the compact topological spaces X_i be given, where i is in I. By Zorn's[7] Lemma 3.2, $\mathcal{C}$ is included in a maximal centered family $\mathcal{M}$ of subsets of X. The finite intersections of all sets belonging to $\mathcal{M}$ form a centered family, too. By the maximality of $\mathcal{M}$, these intersections belong to $\mathcal{M}$. This implies that if the subset $A \subseteq X$ has nonempty intersection with every element of the family $\mathcal{M}$, then the sets belonging to $\mathcal{M}$ together with A form a centered family again, hence, by the maximality of $\mathcal{M}$, such a set A must belong to $\mathcal{M}$, too. For each i in I the images of the sets in $\mathcal{M}$ by the projection p_i of X onto X_i form a centered family $\mathcal{M}_i$ in X_i, and clearly so does the family of their closures. As X_i is compact, the intersection of the closures of the sets belonging to $\mathcal{M}_i$ contains an element x_i. If U is an open set in X_i containing x_i, then the intersection of U with each of the sets in $\mathcal{M}_i$ is nonempty, hence the inverse image of U by p_i has nonempty intersection with every set in $\mathcal{M}$, consequently, this inverse image belongs to $\mathcal{M}$, too. This implies that the intersection of finitely many inverse images of this type belongs to $\mathcal{M}$, hence every open set containing the function f in X defined by $f(i) = x_i$ at the point i in I has a nonempty intersection with all sets in $\mathcal{M}$. In other words, the closure of every set in $\mathcal{M}$ contains f. It follows that the sets belonging to $\mathcal{C}$ have nonempty intersection, which proves that Y is a compact topological space. □

A. N. Tikhonov

M. Zorn

This theorem is a fundamental result on compact topological spaces.

[6]Andrei Nikolaevich Tikhonov, Russian mathematician (1906-1993)
[7]Max August Zorn, German mathematician (1906-1993)

In fact, this theorem makes clear and justifies the given definition of the topology on products of topological spaces.

A topological space, in which every point has a closed compact neighborhood, is called *locally compact.* Obviously, compact topological spaces are locally compact.

In contrast to metric spaces, in topological spaces the points in the closure of a subset are not necessarily identical with the limits of sequences in the subset. To describe closures of sets we need generalized sequences.

Let X be a topological space, and let $(x_i)_{i\in I}$ be a generalized sequence in X. We say that this generalized sequence *converges* to the element a of X, if for every neighborhood U of a there exists an i_0 in I such that for $i_0 \leqslant i$ x_i is in U. It may happen that a generalized sequence converges to several different points, but we shall see that this is excluded in Hausdorff spaces: in a Hausdorff topological space a generalized sequence $(x_i)_{i\in I}$ converges to at most one point, which is called the *limit* of the generalized sequence, and it is denoted by $\lim_i x_i$.

E. H. Moore

The benefit of generalized sequences is shown by the following two theorems, whose analogues are well-known in metric spaces concerning ordinary sequences.

Theorem 3.5. *In a topological space the closure of an arbitrary subset is equal to the set of those points of the space to which generalized sequences belonging to the subset are converging.*

Proof. Let h be a point of the closure of the set H, and let I be the set of all open neighborhoods of the point h. We introduce partial ordering on I as follows: if U, V are neighborhoods of h, then we let $U \leqslant V$, whenever $U \supseteq V$. It is clear that this is a partial ordering on I, which makes I a directed set. As every neighborhood of h contains an element of H, hence, by the Axiom of Choice, there is a function $x : I \to H$ such that $x(U)$ belongs to U whenever U is in I. Then, by the definition $x_U = x(U)$ we obtain a generalized sequence $(x_U)_{U\in I}$ in H, which obviously converges to h.

Conversely, if $(x_i)_{i\in I}$ is a generalized sequence in H converging to h, then obviously every neighborhood of h contains a member of H, hence h belongs to the closure of H. □

Theorem 3.6. *Let X, Y be topological spaces, x a point in X, and $f : X \to Y$ a function. Then f is continuous at x if and only if for each generalized sequence $(x_i)_{i\in I}$ in X converging to x the generalized sequence $\big(f(x_i)\big)_{i\in I}$ converges to $f(x)$.*

Proof. First suppose that f is continuous at x, and the generalized sequence $(x_i)_{i\in I}$ converges to x. Let V be an arbitrary neighborhood of $f(x)$, then there exists a neighborhood U of x such that for each y in U $f(y)$ belongs to V. For the given U there exists an element i_0 in I such that x_i is in U, whenever $i \geqslant i_0$, hence $f(x_i)$ is in V for $i \geqslant i_0$, which means that the generalized sequence $\big(f(x_i)\big)_{i\in I}$ converges to $f(x)$.

Conversely, we suppose that for every generalized sequence $(x_i)_{i\in I}$ in X converging to x the generalized sequence $\big(f(x_i)\big)_{i\in I}$ converges to $f(x)$, but f is not continuous at x. This means that there exists a neighborhood V of $f(x)$ such that every neighborhood U of x contains an element x_U such that $f(x_U)$ is not in V. We see immediately that the generalized sequence $(x_U)_{U\in I}$ obtained in this way on the directed set I, introduced in Theorem 3.5, converges to x, however, the generalized sequence $\big(f(x_U)\big)_{U\in I}$ does not converge to $f(x)$, a contradiction. The theorem is proved. □

3.3 Separation theorems, Uryshon's Lemma

A topological space is said to be T_0*-space*, if given any two points of the space at least one of them has a neighborhood, which does not contain the other. A topological space is said to be T_1*-space*, if every one point set in the space is closed. Obviously, every T_1-space is T_0-space. A topological space is said to be T_2*-space*, or *Hausdorff*[8] space, if for any two points in the space there are disjoint open sets, one containing one of the two points, and the other containing the other point. As it is easy to see, every Hausdorff space is T_1-space. It is also obvious that the product of Hausdorff spaces

[8]Felix Hausdorff, German mathematician (1868-1942)

is a Hausdorff space. A topological space is said to be *regular*, if given any closed set, and a point outside the set there are disjoint open sets, one including the closed set, and the other containing the point.

A topological space is said to be *completely regular*, if its topology can be induced by real valued continuous functions. As it is easy to see, this is the case if and only if given any closed set in the space, and a point outside the set there exists a real valued continuous function vanishing on the closed set, and taking the value 1 at the point.

A topological space is said to be *normal*, if given two disjoint closed sets there are disjoint open sets, one containing one of the closed sets, and the other containing the other.

Theorem 3.7. *(Uryshon's[9] Lemma) The topological space X is normal if and only if for every disjoint closed sets A, B there is a continuous mapping of X into the closed interval $[a, b]$ with $a < b$ such that it is identically a on A, and it is identically b on B.*

Proof. Using the linear transformation $f = (b-a)g + a$ we may suppose that $[a, b] = [0, 1]$. As X is normal, if U is open and H is closed with $H \subseteq U$, then there is an open set V such that $H \subseteq V \subseteq V^{cl} \subseteq U$ holds. We define the sets $U_{2^{-n}k}$ by induction ($n, k \geqslant 0$ are natural numbers, $k \leqslant 2^n$) as follows. Let U_1 be the complement of B, and let U_0 be an open set for which $A \subseteq U_0 \subseteq U_0^{cl} \subseteq U_1$ holds. Suppose that we have defined the sets $U_{2^{-n}k}$ for some n and for each $k \leqslant 2^n$ such that $U_{2^{-n}k}$ is open, and $k < k' \leqslant 2^n$ implies $U_{2^{-n}k}^{cl} \subseteq U_{2^{-n}k'}$. Let for $k < 2^n$ the open set $U_{2^{-(n+1)}(2k+1)}$ have the property

$$U_{2^{-n}k}^{cl} \subseteq U_{2^{-(n+1)}(2k+1)} \subseteq U_{2^{-(n+1)}(2k+1)}^{cl} \subseteq U_{2^{-n}(k+1)} .$$

Clearly, $k < k' \leqslant 2^{n+1}$ implies $U_{2^{-(n+1)}k}^{cl} \subseteq U_{2^{-(n+1)}k'}$. The sets $U_{2^{-n}}$ obviously include A, and they are disjoint to B, moreover, for the numbers $0 \leqslant r < r' \leqslant 1$ of the form $2^{-n}k$ we have $U_r^{cl} \subseteq U_{r'}$. If x belongs to U_1, then let $f(x)$ denote the infimum of those numbers r of the above type, for which x belongs to U_r, further, if x does not belong to U_1, then we let $f(x) = 1$. Clearly, f maps X into $[0, 1]$, it takes 0 at the elements of A, and it takes 1 at the elements of B. On the one hand, given a real number $0 < t \leqslant 1$ the inequality $f(x) < t$ holds exactly for those x in X, which belong to the union of the sets U_r with $r < t$, hence the set of all elements

[9] Pavel Samuilovich Uryshon, Russian mathematician (1898-1924)

x with this property is open. On the other hand, given a real number t the inequality $f(x) > t$ holds exactly for those x in X, which belong to the union of the complements of the sets U_r^{cl} with $r < t$, hence the set of these elements x is an open set, too. It follows that f is continuous.

The converse statement is obvious. □

F. Hausdorff P. S. Urysohn H. F. F. Tietze

Theorem 3.8. *(Tietze's Extension Theorem)*[10] *Every continuous bounded real valued function defined on a closed subset in a normal topological space can be extended to a continuous function on the whole space with the same bounds.*

Proof. Let $[a, b]$ be the smallest closed interval containing the range of the continuous function $f : F \to \mathbb{R}$, where F is a closed subset of the normal topological space X. Obviously, we may suppose that $a < b$. Moreover, applying the linear transformation $f = \frac{1}{2}(b-a)g + \frac{1}{2}(b+a)$ we may also suppose that $[a, b] = [-1, 1]$.

We define the following subsets of F:

$$A_0 = \{x : f(x) \leqslant -\frac{1}{3}\} \text{ and } B_0 = \{x : f(x) \geqslant \frac{1}{3}\}.$$

Then A_0, B_0 are disjoint closed subsets of F, and, as F is closed in X, A_0, B_0 are disjoint closed subsets of X. By Uryshon's Lemma, there is a continuous function $g_0 : X \to [-\frac{1}{3}, \frac{1}{3}]$ such that $g_0 = -\frac{1}{3}$ on A_0 and $g_0 = \frac{1}{3}$ on B_0. Now let $f_0 = f$ and $f_1 = f_0 - g_0$ on F, then $|f_1| \leqslant \frac{2}{3}$. Again, we define the following subsets of F:

$$A_1 = \{x : f_1(x) \leqslant -\frac{1}{3} \cdot \frac{2}{3}\} \text{ and } B_1 = \{x : f_1(x) \geqslant \frac{1}{3} \cdot \frac{2}{3}\},$$

then, in the same way as above, there exists a continuous function

$$g_1 : X \to \left[-\frac{1}{3} \cdot \frac{2}{3}, \frac{1}{3} \cdot \frac{2}{3}\right]$$

[10] Heinrich Franz Friedrich Tietze, Austrian mathematician (1880-1964)

with $g_1 = -\frac{1}{3} \cdot \frac{2}{3}$ on A_1, and $g_1 = \frac{1}{3} \cdot \frac{2}{3}$ on B_1. We next define $f_2 = f_1 - g_1 = f_0 - (g_0 + g_1)$ on F, and observe that $|f_2| \leqslant \left(\frac{2}{3}\right)^2$. By continuing in this manner, we get a sequence $(f_n)_{n\in\mathbb{N}}$ on F satisfying $|f_n| \leqslant \left(\frac{2}{3}\right)^n$, and another sequence $(g_n)_{n\in\mathbb{N}}$ on X satisfying

$$|g_n| \leqslant \frac{1}{3}\left(\frac{2}{3}\right)^n \tag{3.1}$$

with

$$f_n = f_0 - (g_0 + g_1 + \cdots + g_{n-1}) \tag{3.2}$$

on F. Let $f^* = \sum g_n$ on X. By (3.1) this series is uniformly convergent on X, and $|f^*| \leqslant 1$. By the continuity of the functions g_n, the function f^* is continuous. Finally, as $|f_n| \leqslant \left(\frac{2}{3}\right)^n$ on F, it follows from (3.2) that $f^* = f_0 = f$ on F, hence f^* is the desired extension. □

Theorem 3.9. *Every normal Hausdorff space is completely regular.*

Proof. The statement follows immediately when applying Uryshon's Lemma to a closed set and a point outside it. □

Theorem 3.10. *For every point in a Hausdorff space, and a compact subset not containing the point there are disjoint open sets, one containing the point, and the other including the compact set.*

Proof. For each point of the compact set Y there exists an open subset of X, the closure of which does not contain x. There is a finite subfamily of these open sets, which form a covering of Y. Clearly, the intersection of the complements of the closures of these finitely many open sets is an open set disjoint to U and containing x. □

Theorem 3.11. *Closed subsets in a compact topological space are compact.*

Proof. Let K be a closed set in the compact topological space X, and let $(U_i)_{i\in I}$ be an open covering of K. We have to show that the union of a finite number of the U_i's includes K. By adding the open complement K^c of K to the covering $(U_i)_{i\in I}$ we obtain the open covering

$$\{U_i : i \in I\} \cup K^c$$

of the compact set X, which has a finite subcovering of the set $X = K \cup K^c$. If the corresponding finite subfamily does not include K^c, then this finite subfamily is a finite subcovering of the original covering of K, which covers K, and the proof is complete. On the other hand, if the corresponding finite subfamily contains K^c, then omitting it the rest is a part of the original covering, and obviously it is still a covering of K, as the points of K are not covered by K^c. □

Theorem 3.12. *Compact subsets of a Hausdorff topological space are closed.*

Proof. Suppose that the complement K^c of the compact subset K of the Hausdorff topological space X contains a point x, which is a boundary point of K^c, hence it is a boundary point of K, too. The Hausdorff property implies that for each p in K there are open sets U_p, V_p such that p is in U_p, x is in V_p, and $U_p \cap V_p = \varnothing$. The family $(U_p)_{p\in K}$ is an open covering of K, hence there are points $p_1, p_2, \dots, p_n$ in K such that

$$K \subseteq U_{p_1} \cap U_{p_2} \cap \cdots \cap U_{p_n}$$

holds. If $U = U_{p_1} \cap U_{p_2} \cap \cdots \cap U_{p_n}$, and $V = V_{p_1} \cap V_{p_2} \cap \cdots \cap V_{p_n}$, then $K \subseteq U$, V is a neighborhood of x, and $U \cap V = \varnothing$, which contradicts the fact that x is a boundary point of K. □

Theorem 3.13. *Compact Hausdorff space is normal.*

Proof. The disjoint closed sets A, B in the compact Hausdorff topological space X are compact, by Theorem 3.11. By Theorem 3.10, for each element of B there exists an open subset of X containing the point, whose closure is disjoint to A. The union U of finitely many of these open sets includes B. It is obvious that the intersection of the complements of the closures of these finitely many open sets is an open set including A, and is disjoint to the open set U. □

Theorem 3.14. *Continuous image of a compact topological space is compact.*

Proof. Let $f : X \to Y$ be a continuous surjective mapping, and suppose that X is compact, further let $(V_i)_{i\in I}$ be an open covering of Y. Then $\left(f^{-1}(V_i)\right)_{i\in I}$ is an open covering of X, having a finite subfamily $f^{-1}(V_{i_1}), f^{-1}(V_{i_2}), \dots, f^{-1}(V_{i_n})$ whose union is X. Then the union of the finite subfamily $V_{i_1}, V_{i_2}, \dots, V_{i_n}$ of the original covering is Y. □

Theorem 3.15. *Continuous injective mapping on a compact Hausdorff topological space into a Hausdorff topological space is open.*

Proof. Let $f : X \to Y$ be a continuous injective mapping, and let U be an open set on the compact Hausdorff space X. Then U^c is closed, hence compact. It follows, by the previous theorem, that $f(U^c)$ is compact, hence closed. Finally, by injectivity, $f(U) = f(U^c)^c$, which is open. □

3.4 Compactification

An important property of locally compact Hausdorff spaces is that they can be considered as a dense subspace of some compact Hausdorff space – by "adding" a single point.

A *compactification* of the topological space X we mean a compact topological space $\widetilde{X}$, which has a dense subspace homeomorphic to X. For more about compactification see e.g. [Munkres (1975)].

Theorem 3.16. *(One-point compactification) Every noncompact topological space X has a compactification $\widetilde{X}$ such that the complement of the image of X in $\widetilde{X}$ is a singleton. The space $\widetilde{X}$ is Hausdorff if and only if X is a locally compact Hausdorff space.*

Proof. Let ∞ denote an arbitrary set different from each element of X, and let

$$\widetilde{X} = X \cup \{\infty\} .$$

We introduce a topology on $\widetilde{X}$ by the following rule: we call open sets in $\widetilde{X}$ all open subsets of X, further those including ∞ and having a complement, which is a closed, compact subset of X. As closed subsets of compact sets are compact, by Theorem 3.11, it follows that the family of sets defined in this way is closed with respect to forming arbitrary unions and finite intersections. It follows that we obtain a topology on $\widetilde{X}$. It is also easy to see that the restriction of this topology to X coincides with its original topology, that is, a subset in X is open in X if and only if it is open in $\widetilde{X}$. As X is noncompact, every neighborhood of ∞ intersects X, that is, X is dense in $\widetilde{X}$. Finally, every open covering of $\widetilde{X}$ contains a set A_0 which includes ∞, and has a closed, compact complement in X. All the other sets in this covering form an open covering of $\widetilde{X}\backslash A_0$, which contains a finite covering $A_1, A_2, \dots, A_n$. Using this we obtain a finite covering $A_0, A_1, \dots, A_n$ of $\widetilde{X}$, which is a subcovering of the original one, hence $\widetilde{X}$ is compact.

If X is locally compact, and x is a point of it, then let C be compact neighborhood of x in X. Then C is closed, by Theorem 3.12, hence the interior of C and $\widetilde{X}\backslash C$ are disjoint neighborhoods of x and ∞ in $\widetilde{X}$. As X is Hausdorff space, any two points of X can be separated by open sets, and we infer that $\widetilde{X}$ is a Hausdorff space, too.

Conversely, every open subset of a locally compact Hausdorff space is a locally compact space, as it is easy to see. □

3.5 Partition of the unity

The theorem on the partition of the unity is useful when drawing global consequences from local properties of functions (see e.g. [Rudin (1987)]). By a minor modification of the proof of Uryshon's Lemma 3.7, we have the following lemma.

Lemma 3.5.1. *Let X be a locally compact Hausdorff space, U an open subset of X, and K a compact subset of U. Then there is a continuous function $f : X \to \mathbb{R}$ such that it is at least 1 on K, and it is vanishing outside a compact subset of U.*

We shall also need the following result.

Lemma 3.5.2. *Let X be a locally compact Hausdorff space, U an open subset of X, and K a compact subset of U. Then there exists an open set V with compact closure satisfying*

$$K \subseteq V \subseteq V^{cl} \subseteq U\,.$$

Proof. As every point in K has a neighborhood with compact closure and a finite number of them is a covering of K, hence K is contained in an open set W having compact closure. We apply the previous lemma: there is a continuous function $f : X \to \mathbb{R}$ such that it is at least 1 on K and it is vanishing outside the compact subset L of U. Then we let

$$V = \{x : \, f(x) > \frac{1}{2}\} \cap W\,.$$

Clearly, V is open, and $K \subseteq V$. If x is in V^{cl}, then $f(x) \geqslant \frac{1}{2}$, hence x is not in L^c, which implies that x is in U. It follows $V^{cl} \subseteq U$. Finally, obviously $V^{cl} \subseteq W^{cl}$, hence V^{cl} is compact. □

From these results we derive the following theorem.

Theorem 3.17. *(Partition of Unity) Let X be a locally compact Hausdorff space, K a compact subset of X, and $\{V_1, V_2, \ldots, V_n\}$ an open covering of K. Then there are nonnegative continuous functions $h_i : X \to \mathbb{R}$ such that any of them is zero outside a compact subset of some V_i, further*

$$h_1(x) + h_2(x) + \cdots + h_n(x) = 1$$

holds for each point x in K.

Proof. By the previous lemma every point x of K has a neighborhood W_x with compact closure such that W_x^{cl} is a subset of some set V_i. Here obviously i may depend on x. Let the points $x_1, x_2, \dots, x_m$ be given with

$$K \subseteq W_{x_1} \cup W_{x_2} \cup \cdots \cup W_{x_m}.$$

If $1 \leqslant i \leqslant n$, then let H_i denote the union of all sets $W_{x_j}^{cl}$ in V_i. By Lemma 3.5.1, there are continuous functions g_i, which vanish outside some compact set, and g_i takes 1 on H_i, and it takes 0 outside V_i. Let $h_1 = g_1$, further let for $2 \leqslant k \leqslant n$

$$h_k = (1 - g_1)(1 - g_2) \dots (1 - g_{k-1}) g_k.$$

Then h_i is continuous, vanishes outside a compact set, and is zero outside V_i. By induction we have

$$h_1 + h_2 + \cdots + h_n = 1 - (1 - g_1)(1 - g_2) \dots (1 - g_n).$$

As $K \subseteq H_1 \cup H_2 \cup \cdots \cup H_n$, it follows that for each point x in K we have $g_i(x) = 1$ for at least one i, and the statement is proved. □

The couple of functions $h_1, h_2, \dots, h_n$ is called a *partition of unity* subordinate to the covering $\{V_1, V_2, \dots, V_n\}$.

3.6 Connectedness

A topological space is called *connected*, if it is not the union of two nonempty proper open subsets. Obviously, this is equivalent to the property that it is not the union of two nonempty proper closed subsets. Another equivalent is that there is no nonempty proper subset in the space, which is open and closed. Sometimes such subsets are called *clopen*. As in every topological space the empty set and the whole space are the trivial clopen sets, hence connectedness is equivalent to the property that in the space there are no nontrivial clopen subsets. A subset of a topological space is called connected, if it is connected, as a subspace. It is easy to see that the union of a family of connected subsets with nonempty intersection is connected, too. Hence, for a given point in a topological space the union of all connected sets containing this point is connected, and it is called the *component* of the point. As it is easy to see, any set between a connected set and its closure is also connected, which implies that all components are closed sets. The components of all points in a topological space form a partition of the space, and they are called the *components of the space*. Clearly, a topological space is connected if and only if it has exactly one component. A topological space is called *totally disconnected*, if each singleton is a component, or, what is the same, each component is a singleton. In particular, every discrete topological space is totally disconnected. A topological space is called *locally connected*, if every point has a neighborhood base consisting of connected sets.

Theorem 3.18. *The continuous image of a connected topological space is connected.*

Proof. Let $f : X \to Y$ be a continuous surjective mapping, where X, Y are topological spaces and X is connected. If A, B are nonempty disjoint open sets in Y and $A \cup B = Y$, then $f^{-1}(A), f^{-1}(B)$ are nonempty disjoint open sets in X and $f^{-1}(A) \cup f^{-1}(B) = X$. □

Lemma 3.6.1. *The product of finitely many connected topological spaces is connected.*

Proof. It is enough to show the statement for two spaces. Let X, Y be connected topological spaces, further let $p = (x_1, y_1)$ and $q = (x_2, y_2)$ be two arbitrary points in the product space $X \times Y$. Then $\{x_1\} \times Y$ is homeomorphic to Y, and $X \times \{y_2\}$ is homeomorphic to X, hence these sets are connected in $X \times Y$. On the other hand, (x_1, y_2) is a common point of them, hence their union, $(\{x_1\} \times Y) \cup (X \times \{y_2\})$ is connected, too. As this set includes both p and q, we have shown that any two points of the space are included in the same connected set, which means that any two points have the same component. It follows that the product space has a single component, hence it is connected. □

Theorem 3.19. *The product of any family of connected topological spaces is connected.*

Proof. Suppose that the factors X_i, where i is in the set I, of the product space $X = \prod_{i \in I} X_i$ are connected topological spaces, let p be a point in X, and let C denote the component of p. We show that each x in X belongs to the closure of C. As every component is a closed set, this will imply that each point of the space belongs to the component of p, that is, the space has a single component, hence it is connected.

Let U be an element of the neighborhood basis of x, which means, that there are elements $i_1, i_2, \ldots, i_m$ in I, and there are open sets U_{i_k} in X_{i_k} such that

$$U = \{f : f \in X \text{ and } f(i_k) \in U_{i_k},\ k = 1, 2, \ldots, m\},$$

and x_{i_k} belongs to U_{i_k}, whenever $k = 1, 2, \ldots, m$. Then the set

$$H = \{f : f \in X \text{ and } f(i) = p(i) \text{ ha } i \neq i_k,\ k = 1, 2, \ldots, m\}$$

is homeomorphic to the product space $X_{i_1} \times X_{i_2} \times \cdots \times X_{i_m}$, which is, by the previous lemma, connected. Moreover, p belongs to H, hence H is a subset of the component C of p. For each i in I we let $q(i) = p(i)$, if i is different from each i_k, and otherwise let $q(i) = x(i)$ for $i = i_k$ $(k = 1, 2, \ldots, m)$. Then q is a common element of U and H, thus $U \cap C$ is nonempty, which means that every neighborhood of x intersects C, that is, x belongs to the closure of C, which is C itself. Consequently $X = C$, and the theorem is proved. □

3.7 Topological groups

In this section we summarize the basic knowledge on the most important structure in this book: the topological group. For further references the reader should consult with [Pontryagin (1966); Montgomery and Zippin (1974); Bourbaki (1998a); Stroppel (2006); Chandrasekharan (2011)]. In these sections the group operation will be written as multiplication. If A, B are subsetes in a group, then we use the obvious notation $A \cdot B = \{a \cdot b : a \in A, b \in B\}$ and $A^{-1} = \{a^{-1} : a \in A\}$. The subset A is called *symmetric*, if $A^{-1} = A$.

A group G, which is also a topological space, is called a *topological group*, if the group operation and the inversion are continuous. As it is easy to see, every group is a topological group when equipped with the discrete topology. It is also easy to see that in a topological group the left and right translations, and the inversion are homeomorphisms. It follows that for a given element g in the topological group G a family of sets is a neighborhood basis of the point g if and only if the left translates of the sets in the family by the inverse of g form a neighborhood basis of the identity element. The same holds for the right translates.

Theorem 3.20. *Each point in a topological group has a neighborhood basis consisting of open symmetric sets.*

Proof. By our above discussion it is enough to prove this statement for the identity element. For every open neighborhood U of the identity the set $U \cap U^{-1}$ is an open and symmetric neighborhood of the identity, which is contained in U. □

Theorem 3.21. *For every open neighborhood U of the identity e in the topological group G we have the following statements:*

1. *there is an open neighborhood V of e such that $V^2 \subseteq U$;*
2. *there is an open neighborhood V of e such that $V^{-1} \subseteq U$;*
3. *for each x in U there is an open neighborhood V of e such that $xV \subseteq U$.*

Proof. The first statement follows from the continuity of the group multiplication, the second follows from the continuity of the inversion, and the third is a consequence of the fact that U is open. □

It is easy to see that every neighborhood U of the identity e contains an open neighborhood V of e such that $V \cdot V^{-1} \subseteq U$ holds.

Theorem 3.22. *Every T_0-topological group is a Hausdorff space.*

Proof. Let x and y be distinct points in the topological group, and let U be a neighborhood of the identity, for which xU does not include y. We choose an open neighborhood V of the identity with $V \cdot V^{-1} \subseteq U$. Then xV, resp. yV are disjoint open sets not including y, resp. x. □

In this work we shall always assume that all topological groups in question have the T_0-property, that is, given any two distinct elements at least one of them has a neighborhood not including the other. By the previous theorem, this implies that the topological group, as a topological space, is a Hausdorff space. In other words, by a topological group we shall **always** mean a Hausdorff topological group.

Theorem 3.23. *Each point in a topological group has a neighborhood base consisting of closed sets.*

Proof. It is enough to show that every neighborhood U of the identity contains a neighborhood V of the identity, whose closure is in U. Let V be a neighborhood of the identity with $V \cdot V^{-1} \subseteq U$. If x is in the closure of V, then the set xV intersects V, hence x belongs to $V \cdot V^{-1}$, which is a subset of U. □

Theorem 3.24. *Every topological group is regular.*

Proof. If F is a closed set and x is a point outside F, then the complement of F is an open neighborhood of x, which contains an open neighborhood V of x, whose closure is in the complement of F, hence V and the complement of the closure of V are disjoint open sets, V contains x, and the complement of the closure of V contains F. □

Theorem 3.25. *Let A and B be subsets of a topological group. If A is open, then $A \cdot B$ and $B \cdot A$ are open. If A is closed and B is compact, then $A \cdot B$ and $B \cdot A$ are closed. If A and B are compact, then $A \cdot B$ is compact.*

Proof. If A is open, then for each b in B the sets Ab and bA are open, hence their union, $A \cdot B$, and $B \cdot A$ are open, too. Now let A be closed, B compact, and let x_0 be the limit of the generalized sequence $(x_\nu)_{\nu \in \Gamma}$ in $A \cdot B$. It is enough to show that x_0 is in $A \cdot B$. It is obvious that for each ν in Γ we have $x_\nu = y_\nu z_\nu$, where y_ν is in A, z_ν is in B. As B is compact, z_ν has a generalized subsequence $z_{\varphi(\beta)}$ converging to some z_0 in B. The corresponding generalized subsequence $x_{\varphi(\beta)}$ converges to x_0, too, hence, by continuity of multiplication and inversion, the generalized sequence $y_{\varphi(\beta)} = x_{\varphi(\beta)} z_{\varphi(\beta)}^{-1}$ converges to $x_0 z_0^{-1}$, which therefore belongs to A, as A is closed. It follows that $x_0 = (x_0 z_0^{-1}) z_0$ is an element of $A \cdot B$.

If A and B are compact, then, by Tikhonov's Theorem 3.4, so is $A \times B$, hence its image $A \cdot B$ at the continuous mapping $(x, y) \mapsto xy$ is compact, too. □

An isomorphic mapping of a topological group onto another is called *topological isomorphism*, if it is a homeomorphism. In this case the two topological groups are said to be *topologically isomorphic*.

3.8 Topological subgroups, factor groups

It is easy to see that every subgroup of a topological group, as a topological subspace is a topological group, too, which is called a *subgroup* of the topological group. It is also obvious that the closure of any subgroup of a topological group is a subgroup, too.

Theorem 3.26. *A subgroup of a topological group is open if and only if it has an interior point. Every open subgroup is closed.*

Proof. If x is an interior point of the subgroup H, then the identity has a neighborhood U such that $xU \subseteq H$. For each element y in H we have $yU = yx^{-1}xU \subseteq yx^{-1}H = H$, hence H is open.

If H is an open subgroup, then its complement is the union of all cosets of the form xH, where x is in G but not in H. These cosets are also open, and so is there union, hence H is closed. □

Theorem 3.27. *For each open, symmetric neighborhood U of the identity in a topological group $\bigcup_{n=1}^{\infty} U^n$ is an open and closed subgroup.*

Proof. If x is in U^k and y is in U^l, then xy is in U^{kl}, and x^{-1} is in $U^{-1} = U$, hence the above union is a subgroup, which is obviously open, hence, by the previous theorem, it is also closed. □

Theorem 3.28. *If a subgroup of a topological group has closed intersection with a closed neighborhood of the identity, then the subgroup is closed.*

Proof. Let the intersection of the closure of the subgroup H with the closure of the neighborhood U of the identity be closed, further let V be an open and symmetric neighborhood of the identity with $V^2 \subseteq U$. Let x be a point in the closure of H, and $(x_\nu)_{\nu\in\Gamma}$ a generalized sequence in H converging to x. As the closure of H is a subgroup, too, hence x^{-1} is in the closure of H and Vx^{-1} intersects H; let y be an element of this intersection. There is a ν_0 such that for $\nu \geqslant \nu_0$ the point x_ν belongs to xV, hence in this case yx_ν belongs to the set $(Vx^{-1})(xV) = V^2 \subseteq U$, therefore yx_ν is a common element of the closure of U and H. As the generalized sequence $(yx_\nu)_{\nu\in\Gamma}$ converges to yx, hence yx is also a common element of the closure of U and H, which means that $x = y^{-1}yx$ belongs to H. In other words, the closure of H is contained in H, hence H is closed. □

Theorem 3.29. *A subgroup of a topological group is discrete if and only if it has an isolated point. Every discrete subgroup is closed.*

Proof. If x is an isolated point of the subgroup H, then there is a neighborhood U of the identity such that $xU \cap H = \{x\}$. For each y in H we have $(yU) \cap H = (yU) \cap (yx^{-1}H) = yx^{-1}((xU) \cap H) = \{y\}$, hence every point of H is isolated, that is, H is discrete. The converse is obvious.

If H is a discrete subgroup, then let U be a neighborhood of the identity e such that $U \cap H = \{e\}$. By Theorem 3.23, the identity has a neighborhood V with closure in U, hence its intersection with H is $\{e\}$, which is a closed set, as the topological group is Hausdorff. Then, by the previous theorem, our statement follows. □

Theorem 3.30. *A locally compact subgroup of a topological group is closed.*

Proof. Let U be a neighborhood of the identity in G. If the intersection of the closure of U with the locally compact subgroup H is compact in H, then it is compact in G, too, hence it is closed, as G is Hausdorff. Then, by Theorem 3.28, H is closed. □

We say that a topological group is *compactly generated*, if a compact subset generates it.

Theorem 3.31. *In a locally compact topological group the following conditions are equivalent:*

1. *the group is compactly generated;*
2. *the group is generated by an open set with compact closure;*
3. *the group is generated by a neighborhood of the identity with compact closure.*

Proof. It is clear that the third condition implies the second, and the second implies the first. Let K be a compact set generating the group. Obviously, we may assume that the identity is in K. Each point in K has an open neighborhood with compact closure, and all these neighborhoods make an open covering of K. Choosing finitely many of them, whose union also covers K, the union is an open set with compact closure containing K, and obviously generates G. Hence the first condition implies the third. □

Theorem 3.32. *Every compact subset of a locally compact topological group is contained in a compactly generated open subgroup.*

Proof. Let K be the compact subset, and we obviously may assume that it is symmetric, and it includes the identity. Every point of K has an open neighborhood with compact closure, and these sets form an open covering of K. Choosing a finite number of them, whose union U also covers K, we have that U is an open set with compact closure containing K, and the same holds for the open and symmetric set $U \cap U^{-1}$, which generates, by Theorem 3.27, a compactly generated open subgroup. □

It is easy to see that every factor group with respect to a normal subgroup is a topological group, when equipped with the factor topology, which is called the *factor group* of the topological group with respect to the given normal subgroup. The natural homomorphism of a topological group onto any factor group is obviously continuous.

Theorem 3.33. *The natural homomorphism of a topological group onto any factor group is an open mapping.*

Proof. If U is an open set in the topological group G and N is a normal subgroup, then the image of U by the natural homomorphism is

$$UN = \{uN : \ u \in U\},$$

which is open in G/N. □

Theorem 3.34. *The factor group of a topological group with respect to a normal subgroup is a Hausdorff space if and only if the normal subgroup is closed, and it is discrete if and only if the normal subgroup is open.*

Proof. If the normal subgroup N of the topological group G is closed, then its complement is open in G, hence the complement of the singleton $\{N\}$ is open in G/N. It follows that singletons are closed in G/N, hence G/N is a T_1-space. Then G is a T_0-space, and, by Theorem 3.22, it is Hausdorff.

If G/N is a Hausdorff space, then it is a T_1-space, hence $\{N\}$ is closed, and its natural preimage, the set N, is closed in G.

Let N be an open normal subgroup in G. As N is the inverse image by the natural mapping onto G/N of the identity of the factor group, hence the singleton $\{N\}$ is open in G/N, hence G/N is discrete. Conversely, if G/N is discrete, then $\{N\}$ is an open set in G/N, hence its natural inverse image, the set N is open in G. □

Theorem 3.35. *Every factor group of a compact topological group is compact.*

Proof. The factor group is a continuous image of the compact topological group. □

3.9 Topological vector spaces

An important special case of topological groups is presented by those commutative ones, where there is another operation defined on the basic set besides the group operation: multiplication with *scalars*, which are real or complex numbers so that these two operations convert the topological group into a vector space over the *scalar field* in such a way that the multiplication with scalars is a continuous mapping from the topological product of our topological group and the scalar field into the topological group. In this case the topology is called a *vector topology*, and the space equipped with a vector topology is called a *topological vector space*. For basic knowledge on topological vector spaces the reader is referred to [Köthe (1969); Grothendieck (1973); Larsen (1973); Kelley and Namioka (1976); Köthe (1979); Jarchow (1981); Dunford and Schwartz (1988a,b,c); Rudin (1991); Yosida (1995)]. Most of the material in the sections on topological vector spaces we took from [Rudin (1991)]. In topological vector spaces we write the group operation as *addition*, and the identity is called *zero element*, or simply *zero*. Hence a topological vector space is a linear space over the real or complex field equipped with a Hausdorff topology, and the addition and scalar multiplication are continuous. It is easy to see that in a topological vector space multiplications by nonzero scalars are topological isomorphisms of the space, in particular, so is the inversion.

In a vector space there are some subsets of special type, which play a distinguished role. The subset H is called *convex*, if for each x, y in H, and for every scalar $0 \leqslant \lambda \leqslant 1$ the element $\lambda x + (1 - \lambda)y$ is in H. A subset H is called *balanced*, if for each scalar $|\lambda| \leqslant 1$ we have $\lambda H \subseteq H$. The subset H is called *absorbing*, if for each x the element $\lambda^{-1}x$ is in H for some scalar $\lambda > 0$. In a topological space the subset H is called *bounded*, if it is absorbed by every neighborhood of the zero, that is, for each neighborhood U of the zero there is a scalar $\lambda > 0$ such that $\lambda^{-1}H \subseteq U$ holds.

Theorem 3.36. *In a topological vector space every neighborhood of the zero is absorbing.*

Proof. Let U be a neighborhood of zero in the topological vector space X, and let x be a point. As the mapping $\lambda \mapsto \lambda x$ from the scalars into X is continuous, and it is 0 at the zero, hence there is a positive integer n such that $n^{-1}x$ is in U. □

The most important topological vector spaces in analysis are the *locally convex* ones, which means that the zero has a neighborhood basis consisting of convex sets.

Theorem 3.37. *In a topological vector space the zero has a neighborhood basis consisting of balanced open sets. In a locally convex topological vector space the zero has a neighborhood basis consisting of convex balanced open sets.*

Proof. Let U be an open neighborhood of zero. As multiplication by scalars is continuous, hence there is a $\delta > 0$, and there exists an open neighborhood V of zero such that $\lambda V \subseteq U$, whenever $|\lambda| < \delta$. Let W be the union of all sets of the form λV with $|\lambda| < \delta$. Then W is a balanced open neighborhood of zero and $W \subseteq U$. The first statement is proved.

To prove the second statement suppose that U is a convex open neighborhood of zero, and let A denote the intersection of all sets of the form

λU, where $|\lambda| = 1$. Then A is convex. Let $W \subseteq U$ be a balanced open neighborhood of zero, which exists by the first part of our theorem. As W is balanced, hence for $|\lambda| = 1$ we have $\lambda^{-1}W = W$, consequently $W \subseteq \lambda U$, which implies $W \subseteq A$. This means that A is a convex neighborhood of zero, moreover $A \subseteq U$. We show that A is balanced. Let $|\alpha| \leqslant 1$, then there is r and β with $0 \leqslant r \leqslant 1$, $|\beta| = 1$ and $\alpha = r\beta$. We have

$$\alpha A = r\beta A = \bigcap_{|\lambda|=1} r\beta\lambda U = \bigcap_{|\lambda|=1} r\lambda U\,.$$

As λU is convex and it includes zero, it follows $r\lambda U \subseteq \lambda U$. We infer $\alpha A \subseteq A$, which gives that A is balanced. Then the interior of A is a convex balanced open neighborhood of zero, and it is included in U. The theorem is proved. □

One important example for locally convex topological vector space is the *normed space*, which is a vector space equipped with a *norm*, that is, a real valued function $x \mapsto \|x\|$ possessing the following properties: for each elements x, y and scalar λ we have

1. $\|x\| \geqslant 0$ and $\|x\| = 0$ implies $x = 0$;
2. $\|\lambda \cdot x\| = |\lambda|\|x\|$;
3. $\|x + y\| \leqslant \|x\| + \|y\|$.

In a normed space for every real number $r > 0$ the set of all elements with $\|x\| < r$, the so-called *open balls around zero*, where r is the *radius* of the ball, form a basis of a topology, and, as it is easy to see, the space equipped with this topology is a topological vector space. As the open balls are clearly convex sets, this topological space is locally convex. We shall consider every normed space as a topological vector space equipped with this *norm topology*.

A generalization of the norm is the *seminorm*. The function $p : X \to \mathbb{R}$ defined on the vector space X is called a seminorm, if for each elements x, y in X and for every scalar λ we have

1. $p(x) \geqslant 0$;
2. $p(\lambda \cdot x) = |\lambda| p(x)$;
3. $p(x + y) \leqslant p(x) + p(y)$.

Given a seminorm p the *open p-ball* around zero with radius $r > 0$ is defined similarly like above: it is the set of all points x satisfying $p(x) < r$. Given a family $\mathcal{P}$ of seminorms on the vector space X, then X, equipped with the topology, whose sub-basis is formed by all open p-balls around zero corresponding to the seminorms p in the family $\mathcal{P}$, is called a *seminormed space*, and the topology is said to be the topology *induced* by the given family of seminorms. This is always a locally convex vector topology on X, however, it is not necessarily Hausdorff, that is, X is not necessarily a topological vector space with respect to this topology – according to our convention. In order that it is, the family must satisfy an additional condition.

Theorem 3.38. *Let X be a real or complex vector space, and let $\mathcal{P}$ be a family of seminorms on X. The topology induced by the family $\mathcal{P}$ on X is a Hausdorff vector topology if and only if for each $x \neq 0$ in X there is a p in $\mathcal{P}$ such that $p(x) \neq 0$.*

Proof. If the given condition is satisfied, then for each points $x \neq y$ we choose a p with $x - y \neq 0$ with $p(x - y) > 0$, then the sets

$$U = \{z : p(z - x) < \frac{1}{2}p(y - x)\} \text{ and } V = \{z : p(z - y) < \frac{1}{2}p(y - x)\}$$

are disjoint and open, further x is in U, and y is in V. The converse statement is obvious. □

If the family of seminorms satisfies the condition given in this theorem, then it is called a *separating family.* Hence a vector space equipped with the topology induced by a family of seminorms is a locally convex topological vector space if and only if the family is separating.

We note that given a family of seminorms on a vector space, then obviously all seminorms in this family are continuous with respect to the induced topology.

3.10 The Minkowski functional

Let X be a topological vector space and A a convex absorbing set in X. The *Minkowski*[11] *functional* of A is defined as follows: for each x in X we let

$$\mu_A(x) = \inf\{t > 0 : t^{-1}x \in A\}.$$

As A is absorbing, by Theorem 3.36, the Minkowski functional has finite values. We have the following theorem.

[11]Hermann Minkowski, German mathematician (1864-1909)

Theorem 3.39. *Let X be a topological vector space, and let A be an absorbing convex subset of X. Then we have*

1. $\mu_A(x+y) \leqslant \mu_A(x) + \mu_A(y)$, *whenever x, y are in X;*
2. $\mu_A(tx) = t\mu_A(x)$, *whenever x is in X, and $t \geqslant 0$;*
3. *if A is balanced, then μ_A is a seminorm;*
4. *if $B = \{x : \mu_A(x) < 1\}$ and $C = \{x : \mu_A(x) \leqslant 1\}$, then $B \subseteq A \subseteq C$, and $\mu_B = \mu_A = \mu_C$.*

We note that if A is convex and absorbing, then it is easy to see that so are B and C.

Proof. For each x in X let $H_A = \{t > 0 : t^{-1}x \in A\}$. As A obviously contains the zero element, and it is absorbing, it follows that if $s > t$ and t is in $H_A(x)$, then also s is in $H_A(x)$. This means that $H_A(x)$ is a half-line with left endpoint $\mu_A(X)$.

Suppose that $\mu_A(x) < s$, $\mu_A(y) < t$, and we let $u = s + t$. Then x is in $s^{-1}A$, and y is in $t^{-1}A$. As A is convex, hence

$$u^{-1}(x+y) = \frac{s}{u}(s^{-1}x) + \frac{t}{u}(t^{-1}y),$$

which belongs to A, which implies $\mu_A(x+y) < u$. The second and third statements are obvious. Finally, if x is in B, then $\mu_A(x) < 1$, which gives that 1 is in $H_A(x)$, hence x is an element of A. On the other hand, if x is in A, then $\mu_A(x) \leqslant 1$. This implies that $B \subseteq A \subseteq C$, consequently, for each x in X we have

$$H_B(x) \subseteq H_A(x) \subseteq H_C(x),$$

which implies

$$\mu_C(x) \leqslant \mu_A(x) \leqslant \mu_B(x).$$

To prove the converse inequality suppose that $\mu_C(x) < s < t < \mu_B(x)$. Then $s^{-1}x$ belongs to C, hence $\mu_C(s^{-1}x) \leqslant 1$, and this implies

$$\mu_A(t^{-1}x) \leqslant \frac{s}{t} \leqslant 1.$$

We have that $t^{-1}x$ is in B, $\mu_B(t^{-1}x) \leqslant 1$, and $\mu_B(x) \leqslant t$, which is impossible, hence $\mu_C(x) = \mu_B(x)$. □

Theorem 3.40. *Let X be a locally convex topological vector space, and let $\mathcal{B}$ denote a neighborhood basis of zero consisting of convex balanced open sets. Then the family of Minkowski functionals of the sets in $\mathcal{B}$ form a separating family of continuous seminorms.*

Proof. As every element of $\mathcal{B}$ is convex, balanced and absorbing by Theorem 3.36, hence, by the previous theorem, the members of the family $\mathcal{P}$ in question are seminorms. If $x \neq 0$, then x does not belong to some element V of $\mathcal{B}$, and hence $\mu_V(x) \geqslant 1$, that is, the family $\mathcal{P}$ is separating. If x belongs to V in $\mathcal{B}$, then there is a $t > 1$ such that tx is in V, too, as V is open. This implies $\mu_V < 1$ on V, hence if $\varepsilon > 0$ and $x - y$ is in εV, then $\mu_V(x - y) < \varepsilon$. On the other hand,

$$\mu_V(x) \leqslant \mu_V(y) + \mu_V(x - y) \quad \mu_V(y) \leqslant \mu_V(x) + \mu_V(x - y)$$

by the subadditivity of the seminorm. Hence if $\varepsilon > 0$ is given, and $x - y$ is in εV, then we have

$$|\mu_V(x) - \mu_V(y)| \leqslant \mu_V(x - y) < \varepsilon ,$$

that is, μ_V continuous. □

The following theorem shows that the topology of exactly those topological vector spaces can be induced by a separating family of seminorms, which are locally convex.

Theorem 3.41. *(Minkowski) A vector space with a topology on it is a locally convex topological vector space if and only if the topology is induced by a separating family of seminorms.*

Proof. The sufficiency of the condition is obvious: we have seen in Theorem 3.38, and in the remarks before and after it that a vector space, equipped with the topology induced by a separating family of seminorms, is a locally convex topological vector space. Conversely, let X be a locally convex topological vector space, and let $\mathcal{B}$ be a neighborhood of zero consisting of convex, balanced, and open sets. The existence of such a neighborhood basis is guaranteed by Theorem 3.37. Further, by Theorem 3.40, the family of Minkowski functionals of the sets belonging to $\mathcal{B}$ is a separating family of continuous seminorms. Let τ_1 denote the topology induced by this family, and let τ be the original topology of the space. As every Minkowski functional is continuous, hence we have $\tau_1 \subseteq \tau$. Conversely, if W belongs to $\mathcal{B}$, then, by $W = \{x : \mu_W(x) < 1\}$, W is the open μ_W-ball of zero of radius 1, hence it belongs to τ_1, which means $\tau \subseteq \tau_1$, and the theorem is proved. □

H. Minkowski

3.11 Conjugate spaces

By a *linear functional* on the topological vector space X we mean a continuous homomorphism of X into the topological vector space of the scalars. It is easy to see that a linear mapping into the vector space of the scalars is continuous, that is, a linear functional if and only if its restriction to bounded sets is bounded. The set of all linear functionals equipped with the pointwise linear operations is obviously a linear space with respect to the same scalar field. This linear space is denoted by X^*, and it is called the *dual*, or the *conjugate* of X. This space can be equipped with a number of different vector topologies. One of them is the topology of bounded convergence. Let B be a bounded subset of X, and for the linear functional f we let

$$q_B(f) = \sup_{x \in B} |(f(x)| \,.$$

As it is easy to see, $\mathcal{Q} = \{q_B\}$ is a separating family of seminorms on X^*, if B ranges over all bounded subsets of X. The topology induced by the family of seminorms $\mathcal{Q}$ on X^* is called *the topology of bounded convergence*. Thus we get a locally convex topological vector space structure on X^*, and it is easy to check that a generalized sequence of linear functionals is convergent in this topology if and only if the generalized sequence of these functions is uniformly convergent over every bounded subset of X.

Obviously, each finite set of the topological vector space X is bounded. If in the previous definition B ranges over all finite subsets instead of all bounded sets, then we still obtain a separating family of seminorms, and the induced topology is maybe the most important one on dual spaces: this is called the *weak*-topology* of X^*. As uniform convergence on finite sets coincides with pointwise convergence, the weak*-topology is usually called the *topology of pointwise convergence.* The weak*-topology is induced by the following family of seminorms: for each x in X we let $p_x(f) = |f(x)|$, then the topology induced by the family $(p_x)_{x\in X}$ on X^* is – as it is easy to see – the weak*-topology. Clearly, the weak*-topology is, in general, strictly weaker than the topology of bounded convergence.

Suppose now that X is a normed space. Then we can introduce a norm on its conjugate in the following natural way. Let f be a linear functional on X, then f is bounded on the *closed unit ball* of X, which is the set of all elements with norm not greater than 1. Let

$$\|f\| = \sup_{\|x\|\leqslant 1} |f(x)|\,.$$

It is easy to check that $\|f\|$ is a norm on X^*, which is, equipped with this norm, a Banach space. The topology arising in this way is called the *norm topology* of X^*, which is obviously stronger than the topology of bounded convergence, hence it is also stronger than the weak*-topology. In case of normed spaces the *closed unit ball of the conjugate space* is meant to be the set of all linear functionals having norm not greater than 1.

The most important property of the weak*-topology is expressed in the following result.

Theorem 3.42. *(Banach–Alaoglu[12]) In the conjugate space of each normed space the closed unit ball is weak*-compact.*

Proof. For each x in X let I_x denote set of scalars λ with $|\lambda| \leqslant \|x\|$. This means that in case of real normed space I_x is the closed interval $[-\|x\|, \|x\|]$, while in the complex case it is the closed unit disk of radius $\|x\|$ centered at the origin, hence in both cases I_x is a compact topological space. It is clear that the closed unit circle S of the space X^* is a subset of the product space $\prod_{x\in X} I_x$, and it is also easy to see that on S the product topology coincides with the weak*-topology, as both of them coincide with the topology of pointwise convergence of functions. By Tikhonov's Theorem 3.4, it follows that S is a subset of a compact Hausdorff space, and if we show that S is closed in the weak*-topology, then we are ready with the proof of the

[12]Leonidas Alaoglu, Canadian-American mathematician (1914-1981)

theorem. Let $(f_i)_{i\in I}$ be a generalized sequence of linear functionals on the directed set I, and suppose that $\|f_i\| \leqslant 1$ holds for each i in I. If this generalized sequence converges in S to the element f, then the generalized sequence $\big(f_i(x)\big)_{i\in I}$ converges to $f(x)$ for each x in X. As we have

$$f_i(\lambda x + \mu y) = \lambda f_i(x) + \mu f_i(y)\,,$$

whenever x, y are in X, and λ, μ are scalars, hence, by pointwise convergence, f satisfies the same equality, that is, f is a linear mapping of X into the vector space of scalars. Further, as $|f_i(x)| \leqslant \|x\|$ holds for each x in X and i in I, it follows $|f(x)| \leqslant \|x\|$, whenever x is in X. This means that f is a linear functional of norm at most 1, that is, it belongs to S, what was to be proved.

Theorem 3.43. *Let X be a topological vector space over $\mathbb{K} = \mathbb{R}$ or $\mathbb{C}$. Then for each linear functional Φ on X^*, equipped with the weak*-topology, there exists an element x in X such that*

$$\Phi(f) = f(x)\,, \tag{3.3}$$

whenever f is in X^*.

Proof. As Φ is continuous and $\Phi(0) = 0$, there exists a neighborhood of zero in X^* such that $|\Phi(f)| < 1$, whenever f is in V. By the definition of weak*-topology this means that there is an $\varepsilon > 0$, and there are elements $x_1, x_2, \dots, x_n$ in X such that for each f with $|f(x_i)| < \varepsilon$ $(i = 1, 2, \dots, n)$ we have $|\Phi(f)| < 1$. Let M denote the maximum of the numbers $|f(x_i)|$, $i = 1, 2, \dots, n$, and let f be arbitrary in X^*. We define for $k = 1, 2, \dots$

$$f_k = \varepsilon\Big(M + \frac{1}{k}\Big)^{-1} f\,.$$

It follows

$$|f_k(x_i)| = |\varepsilon\Big(M + \frac{1}{k}\Big)^{-1} f(x_i)| \leqslant \frac{\varepsilon M}{M + \frac{1}{k}} < \varepsilon$$

for $i = 1, 2, \dots, n$ and $k = 1, 2, \dots$ This means $|\Phi(f_k)| < 1$. From this we infer

$$\varepsilon\Big(M + \frac{1}{k}\Big)^{-1} |\Phi(f)| < 1 \text{ for } k = 1, 2, \dots\,,$$

which implies

$$|\Phi(f)| < \frac{1}{\varepsilon}\Big(M + \frac{1}{k}\Big)\,,$$

hence we conclude

$$|\Phi(f)| \leqslant \frac{1}{\varepsilon} M . \tag{3.4}$$

We define the linear mapping $F : X^* \to \mathbb{K}^n$ by

$$F(f) = \big(f(x_1), f(x_2), \ldots, f(x_n)\big)$$

for each f in X^*. Then (3.4) implies $\operatorname{Ker} F \subseteq \operatorname{Ker} \Phi$, which means that there is a linear mapping $\sigma : \mathbb{K}^n \to \mathbb{K}$ with $\Phi = \sigma \circ F$. We infer that there are numbers $\lambda_1, \lambda_2, \ldots, \lambda_n$ in $\mathbb{K}$ such that

$$\Phi(f) = \lambda_1 f(x_1) + \lambda_2 f(x_2) + \cdots + \lambda_n f(x_n) = f\big(\sum_{i=1}^{n} \lambda_i x_i\big) ,$$

and the proof is complete. □

3.12 The Hahn–Banach Theorem

This section is devoted to the study of one of the principal theorems of linear analysis. This theorem has a number of different formulations, and here we discuss those, which are the most important ones for our purposes (see e.g. [Rudin (1991)]).

First we state a useful theorem on the continuity of linear mappings on topological vector spaces.

Theorem 3.44. *Let f be a nonzero linear mapping on the topological vector space X into the vector space of scalars. Then the following statements are equivalent:*

1. *f is continuous at some point.*
2. *f is a linear functional.*
3. $\operatorname{Ker} f$ *is closed.*
4. $\operatorname{Ker} f$ *is not dense.*
5. *f is bounded in some neighborhood of zero.*

Proof. Let f be continuous at x_0, and let $(y_i)_{i\in I}$ be a generalized sequence converging to y. Then the generalized sequence $(x_0 + y - y_i)_{i\in I}$ converges to x_0, hence

$$\lim_i [f(y_i) - f(y)] = \lim_i [f(x_0 + y_i - y) - f(x_0)] = 0 ,$$

that is $\big(f(y_i)\big)$ converges to $f(y)$, hence, by Theorem 3.6, f is continuous, hence it is a linear functional.

As $\{0\}$ is closed in the space of scalars, and f is continuous, hence the second condition implies the third one.

If $\operatorname{Ker} f$ is closed, then, by $\operatorname{Ker} f \neq X$, the third condition implies the fourth one.

Suppose that $\operatorname{Ker} f$ is not dense in X. Then there is a point x in X and a balanced neighborhood of zero V such that $x + V$ does not intersect $\operatorname{Ker} f$. As obviously $f(V)$ is balanced in the scalar field, hence it is either bounded, or the whole space. In the latter case there is a y in V such that $f(y) = -f(x)$, which implies that $x + y$ is in $x + V$ and in $\operatorname{Ker} f$, a contradiction. Hence $f(V)$ is a bounded set, which shows that the fourth condition implies the fifth one.

Finally, assume that $|f(x)| < M$ for each x in a neighborhood V of zero. Let $\varepsilon > 0$ and $W = \varepsilon/M \cdot V$. If x is in W, then $M \cdot x/\varepsilon$ is in V, hence

$$|f(x)| = \varepsilon/M \cdot |f(M \cdot x/\varepsilon)| < \varepsilon,$$

and f is continuous at zero. The proof is complete. □

Theorem 3.45. *Let X be a real vector space, M a subspace, and $p : X \to \mathbb{R}$ a function satisfying*

$$p(x + y) \leqslant p(x) + p(y) \quad \text{and} \quad p(tx) = tp(x) \tag{3.5}$$

whenever x, y are in X and $t \geqslant 0$. If $f : M \to \mathbb{R}$ is a linear mapping, satisfying $f(x) \leqslant p(x)$ for each x in M, then there is a linear mapping $\Lambda : X \to \mathbb{R}$ on X such that $\Lambda(x) = f(x)$ for x in M. Moreover, we have $-p(-x) \leqslant \Lambda(x) \leqslant p(x)$ for x in X.

Proof. Supposing $M \neq X$ let x_1 be an element outside M. First we extend f to the subspace

$$M_1 = \{x + t\,x_1 : x \in M,\, t \in \mathbb{R}\}$$

in the following manner. As

$$f(x) + f(y) = f(x + y) \leqslant p(x + y) \leqslant p(x) + p(y) \leqslant p(x - x_1) + p(x_1 + y),$$

hence

$$f(x) - p(x - x_1) \leqslant p(x_1 + y) - f(y)$$

holds for each x, y in M. Let α denote the supremum of the set of values on the left side, whenever x ranges over M. It follows

$$f(x) - \alpha \leqslant p(x - x_1) \quad \text{and} \quad f(y) + \alpha \leqslant p(x_1 + y) \tag{3.6}$$

for each x, y in M. Now we let

$$f_1(x + tx_1) = f(x) + t\alpha\,, \tag{3.7}$$

whenever x is in M, and t is a real number. Then $f_1 : M_1 \to \mathbb{R}$ is a linear mapping, and we have $f_1 = f$ on M. We note that equation (3.7) defines f_1 properly on M_1, because it is easy to see that the representation of the elements of M_1 in the form $x + tx_1$ is unique.

If $t > 0$, and in the first equation of (3.6) we replace x by $t^{-1}x$, further, in the second equation y by $t^{-1}y$, then we multiply the new equations by t, we immediately get $f_1 \leqslant p$ on M_1.

In the second part of the proof we apply Zorn's Lemma 3.2. All linear extensions of f to subspaces containing M, which also satisfy the additional inequality on p, obviously form a partially ordered set, if the partial ordering is defined in the following way: if f', and f'' are extensions to M', respectively to M'', then we let $(f', M') \leqslant (f'', M'')$, whenever $M' \subseteq M''$, and $f'' = f'$ on M'. It is easy to see that in this partially ordered set every chain is bounded from above. Indeed, given a chain we take the extension, whose domain is the union of the extensions in the chain and the extension on this union is defined in the obvious way. Hence, by Zorn's Lemma 3.2, f has a maximal extension, and its domain is necessarily X, by the first part of this proof: otherwise it would have an extension to a strictly larger subspace, which is impossible by maximality.

Observe that we have

$$-p(-x) \leqslant -\Lambda(-x) = \Lambda(x)\,,$$

for each x in X, hence the theorem is proved. □

Theorem 3.46. *Let M be a subspace of the vector space X, f a linear mapping of M into the vector space of the scalars, and let p be a seminorm on X satisfying*

$$|f(x)| \leqslant p(x)\,,$$

whenever x is in M. Then there exists a linear mapping Λ of X into the vector space of the scalars with $\Lambda = f$ on M such that

$$|\Lambda(x)| \leqslant p(x)\,,$$

whenever x is in X.

Proof. If the scalar field is $\mathbb{R}$, then the statement follows from the previous theorem, as in that case $p(-x) = p(x)$.

Let X be a complex vector space, and let u, v denote the real, resp. the imaginary part of f:

$$u(x) = \operatorname{Re} f(x), \quad v(x) = \operatorname{Im} f(x)$$

whenever x is in M. Then u has a linear extension U onto X by the previous theorem such that $U \leqslant p$ on X. We let

$$\Lambda(x) = U(x) - iU(ix), \tag{3.8}$$

whenever x is in X It is easy to see that $\Lambda : X \to \mathbb{C}$ is a linear mapping. For each x in M we have

$$iu(x) - v(x) = i(u(x) + iv(x)) = if(x) = f(ix) = u(ix) + iv(ix),$$

which implies $v(x) = -u(ix)$ and $u(x) = v(ix)$, whenever x is in M. If x is in M, then ix is in M, too, hence

$$\Lambda(x) = U(x) - iU(ix) = u(x) - iu(ix) = u(x) + iv(x) = f(x),$$

that is, Λ is an extension of f. Finally, for each x in X there is a complex number α with $|\alpha| = 1$ and $|\Lambda(x)| = \alpha\Lambda(x)$, hence

$$|\Lambda(x)| = \alpha\Lambda(x) = \Lambda(\alpha x) = U(\alpha x) \leqslant p(\alpha x) = p(x).$$

□

Theorem 3.47. *If X is a normed space, then for each x_0 in X there exists a linear functional f on X such that $f(x_0) = \|x_0\|$, and $|f(x)| \leqslant \|x\|$, whenever x is in X.*

Proof. If $x_0 = 0$, then we let $f = 0$, otherwise we define $p(x) = \|x\|$ for each x in X. Let M denote the one dimensional subspace generated by x_0, on which we let $f(tx_0) = tx_0$. Finally, we apply the previous theorem. □

For the next result we need the following lemma.

Lemma 3.12.1. *Let K, C be disjoint sets in the topological vector space X, with K is compact and C is closed. Then there is an open neighborhood V of zero such that*

$$(K + V) \cap (C + V) = \varnothing.$$

Proof. If $K = \varnothing$, then $K + V = \varnothing$, and there is nothing to prove. Otherwise let x be a point in K. As C is closed and it does not contain x, hence, by applying the first statement in Theorem 3.21 repeatedly, we obtain a symmetric open neighborhood V_x of zero such that the set $x + V_x + V_x + V_x$ does not intersect C, whence, by the symmetry of V_x, the set $x + V_x + V_x$ does not intersect $C + V_x$. As K is compact, there are elements $x_1, x_2, \ldots, x_n$ in it such that the union of the sets $x_i + V_{x_i}$ contains K. We let

$$V = \bigcap_{i=1}^{n} V_{x_i},$$

then it follows

$$K + V \subseteq \bigcup_{i=1}^{n} (x_i + V_{x_i} + V) \subseteq \bigcup_{i=1}^{n} (x_i + V_{x_i} + V_{x_i}),$$

and none of these sets intersects $C + V$. The proof is complete. □

Theorem 3.48. *(Hahn–Banach Separation Theorem) Let A, B be non-empty, disjoint convex subsets in the topological vector space X. Then the following statements hold true.*

1. *If A is open, then there is a linear functional Λ on X and a real number γ such that if x is in A, and y is in B, then*

$$\operatorname{Re}\Lambda(x) < \gamma < \operatorname{Re}\Lambda(y).$$

2. *If X is locally convex, A is compact, and B is closed, then there is a linear functional Λ on X, and there are real numbers γ_1, γ_2 such that if x is in A, and y is in B, then*

$$\operatorname{Re}\Lambda(x) < \gamma_1 < \gamma_2 < \operatorname{Re}\Lambda(y).$$

Proof. First of all we note that it is enough to prove this theorem for real topological vector spaces. Indeed, if we have proved it for those, and X is a complex vector space, then, considering X as a real vector space, there exists a real linear functional U with the desired property, and we let Λ be the functional defined in Theorem 3.46 by equation (3.8).

So let X be a real vector space, a_0 an element of A, b_0 an element of B, further let $x_0 = b_0 - a_0$ and $C = A - B + x_0$. Then C is a convex neighborhood of zero in X. By Theorem 3.39, the Minkowski functional μ_C of the set C satisfies the condition (3.5) in Theorem 3.45. As A and B are disjoint, hence x_0 is not in C, which implies $\mu_C(x_0) \geqslant 1$.

Let $f(tx_0) = t$ on the one dimensional subspace M generated by x_0. Then for each $t \geqslant 0$ we have

$$f(tx_0) = t \leqslant t\mu_C(x_0) = \mu_C(tx_0)\,,$$

and for $t < 0$ we have

$$f(tx_0) = t < 0 \leqslant \mu_C(tx_0)\,.$$

By Theorem 3.45, f has an extension on X to a linear mapping Λ, which also satisfies $\Lambda \leqslant \mu_C$ on the whole X. In particular, $\Lambda \leqslant 1$ on C, hence we have $\Lambda \geqslant -1$ on $-C$, which implies that $|\Lambda| \leqslant 1$ on the neighborhood $V = C \cap (-C)$ of zero. If $\varepsilon > 0$, then $|\Lambda| \leqslant \varepsilon$ on the neighborhood εV of zero, hence Λ is continuous at zero. If x_0 is arbitrary in X, and x is in $x_0 + \varepsilon V$, then

$$|\Lambda(x) - \Lambda(x_0)| = |\Lambda(x - x_0)| \leqslant \varepsilon\,,$$

consequently, Λ is continuous, hence it is a linear functional. If a is in A, and b is in B, then

$$\Lambda(a) - \Lambda(b) + 1 = \Lambda(a - b + x_0) \leqslant \mu_C(a - b + x_0) < 1\,,$$

because $\Lambda(x_0) = f(x_0) = 1$, $a-b+x_0$ belongs to C, and C is open. It follows $\Lambda(a) < \Lambda(b)$. As $\Lambda(A)$ and $\Lambda(B)$ are convex sets on the real line, they are necessarily disjoint intervals, further each element of $\Lambda(A)$ is smaller than every element of $\Lambda(B)$. On the other hand, Λ is an open mapping. Indeed, let y_0 be an element of X with $\Lambda(y_0) = 1$. As Λ is nonzero, such a y_0 does exist. If G is an open set in X and g belongs to it, then there is a $\delta > 0$ such that $|t| < \delta$ implies that $g + ty_0$ is in G, too, hence, by $\Lambda(g + ty_0) = \Lambda(g) + t$, we have that the set $\Lambda(G)$ includes the open interval $]\Lambda(g) - \delta, \Lambda(g) + \delta[$, too. As A is open, this means that $\Lambda(A)$ is open, and γ can be taken as the right endpoint of the open interval $\Lambda(A)$. The first statement is proved.

To prove the second statement we use Lemma 3.12.1, which implies the existence of a convex open neighborhood V of zero such that $A + V$ does not intersect B. Now we can apply what we have proved in the first part using $A + V$ for A to obtain that there exists a linear functional Λ on X such that $\Lambda(A + V)$ and $\Lambda(B)$ are disjoint convex sets, $\Lambda(A + V)$ is open, and it lies on the left of $\Lambda(B)$. As $\Lambda(A)$ is a compact subset of $\Lambda(A + V)$, hence any points $\gamma_1 < \gamma_2$ in the nonempty open interval, determined by the supremum of $\Lambda(A)$ and the right endpoint of $\Lambda(A + V)$, satisfy our requirements. □

Theorem 3.49. *The conjugate of a locally convex topological vector space is a separating family.*

Proof. If $x \neq y$, then we apply the second part of the previous theorem for the sets $A = \{x\}$ and $B = \{y\}$. $\square$

Theorem 3.50. *If M is a subspace in the locally convex topological vector space X, and x_0 is not in the closure of M, then there is a linear functional Λ on X, which vanishes on M and $\Lambda(x_0) = 1$.*

Proof. In the second part of Theorem 3.48 we let $A = \{x_0\}$ and $B = M^{cl}$. Then $\Lambda(x_0)$ is not in $\Lambda(M)$, which means that the latter is a proper subspace in the vector space of the scalars, hence it is $\{0\}$. The desired functional is obtained then through division of Λ by $\Lambda(x_0)$. $\square$

Theorem 3.51. *(Hahn–Banach Extension Theorem) Every linear functional of a linear subspace in a locally convex topological vector space extends to a linear functional of the whole space.*

Proof. Obviously, we may suppose that f is not the zero functional of the subspace M. Let M_0 denote the null space of f: $M_0 = \{x : f(x) = 0\}$, and we choose an x_0 with $f(x_0) = 1$. As f is continuous, hence x_0 does not belong to the M-closure of M_0, hence it does not belong to its X-closure, too. By Theorem 3.50, there exists a linear functional Λ on X, which vanishes on M_0 and $\Lambda(x_0) = 1$. If x is in M, then $x - f(x)x_0$ belongs to M_0, as $f(x_0) = 1$. It follows

$$\Lambda(x) - f(x) = \Lambda(x) - f(x)\Lambda(x_0) = \Lambda\big(x - f(x)x_0\big) = 0\,,$$

hence $\Lambda = f$ on M. $\square$

We note that each of the theorems in this section may be called "Hahn–Banach Theorem": this name usually refers to a group of similar separation and extension theorems.

3.13 The Stone–Weierstrass Theorem

The Weierstrass approximation theorem states that every continuous function defined on a closed interval can be uniformly approximated as closely as desired by a polynomial function. This classical theorem is the root of those general approximation theorems, which belong to the family of Stone–Weierstrass theorems. In this section we study these results. For more references the reader is referred to [Rudin (1964, 1991)].

Let X be a compact Hausdorff space, and let $\mathcal{C}(X)$ denote the set of all continuous complex valued functions on X. The set $\mathcal{C}(X)$, equipped with the pointwise addition of functions and the pointwise multiplication of functions by scalars, is a complex linear space. Moreover, equipped with the pointwise multiplication of functions, the space $\mathcal{C}(X)$ is a commutative complex algebra with unit. Equipped with complex conjugation the space $\mathcal{C}(X)$ is a commutative $*$-algebra with unit. Obviously, the functions in $\mathcal{C}(X)$ are bounded. As it is easy to see, the formula

$$\|f\|_\infty = \sup_{x \in X} |f(x)|$$

defines a norm on the space $\mathcal{C}(X)$. It is easy to see that equipped with this norm $\mathcal{C}(X)$ is a Banach space, moreover, together with the multiplication and complex conjugation, it is a commutative B^*-algebra with unit. In the space $\mathcal{C}(X)$ all real valued functions form a closed subspace, which is denoted by $C_R(X)$.

A subset in $\mathcal{C}(X)$ is called *separating*, if for every two points of the space there is a function in the subset, which takes different values at those points.

First we discuss the real version of the Stone[13]–Weierstrass[14] Theorem.

Theorem 3.52. *(Stone–Weierstrass Theorem, real version) If A is a closed separating subspace in the space $C_R(X)$ of all real valued continuous functions on the compact Hausdorff space X, which contains the constant 1 function, and also the absolute value of each element of it, then $A = C_R(X)$.*

Proof. If f and g are in A, then the functions $f \vee g$ and $f \wedge g$, defined by $(f \vee g)(x) = \max(f(x), g(x))$ and $(f \wedge g)(x) = \min(f(x), g(x))$ at the points x in X, belong to A, too, as $f \vee g = 2^{-1}(f + g + |f - g|)$ and $f \wedge g = 2^{-1}(f + g - |f - g|)$ holds. Hence, if $f_1, f_2, \ldots, f_n$ belong to A ($n \geqslant 2$ is an integer), then we have that

$$f_1 \vee f_2 \vee \cdots \vee f_n = (f_1 \vee f_2 \vee \cdots \vee f_{n-1}) \vee f_n \,,$$

and

$$f_1 \wedge f_2 \wedge \cdots \wedge f_n = (f_1 \wedge f_2 \wedge \cdots \wedge f_{n-1}) \wedge f_n$$

[13] Marshall Harvey Stone, American mathematician (1903-1989)
[14] Karl Theodor Wilhelm Weierstrass, German mathematician (1815-1897)

belong to A, too.

For each $y \neq x$ in X and for arbitrary real numbers $\alpha \neq \beta$ there is an f in A such that $f(x) = \alpha$ and $f(y) = \beta$. Indeed, if the function g in A satisfies $g(x) \neq g(y)$, then there are real numbers γ and δ with the property that the function $f = \gamma g + \delta$ satisfies $f(x) = \alpha$ and $f(y) = \beta$.

Let h be arbitrary in $\mathcal{C}_R(X)$, and for each x, y in X let $f_{x,y}$ be a function in A with $f_{x,y}(x) = h(x)$ and $f_{x,y}(y) = h(y)$. For a given $\varepsilon > 0$ all points z in X satisfying $f_{x,y}(z) < h(z) + \varepsilon$ form an open set $U_{x,y}$. Whenever x is in X the union of all sets of the form $U_{x,y}$ with y in X covers the compact space X, as y is in $U_{x,y}$. It follows that there are elements $y_1, y_2, \ldots, y_n$ in X (n is a positive integer) such that $U_{x,y_1} \cup U_{x,y_2} \cup \cdots \cup U_{x,y_n} = X$. The function

$$f_x = f_{x,y_1} \wedge f_{x,y_2} \wedge \cdots \wedge f_{x,y_n}$$

satisfies $f_x(z) < h(z) + \varepsilon$ for each z in X, moreover $f_x(x) = h(x)$. All elements z in X satisfying $f_x(z) > h(z) - \varepsilon$ form an open set U_x. The sets U_x cover X, as x obviously belongs to U_x, hence there are elements $x_1, x_2, \ldots, x_k$ in X (k is a positive integer) such that $U_{x_1} \cup U_{x_2} \cup \cdots \cup U_{x_k} = X$. Finally, the function

$$f = f_{x_1} \vee f_{x_2} \vee \cdots \vee f_{x_k}$$

obviously satisfies $f(z) < h(z) + \varepsilon$ and $f(z) > h(z) - \varepsilon$, whenever z is in X. As ε is arbitrary and A is closed, the function h is in A, consequently we infer $A = \mathcal{C}(X)$. □

A linear subspace of the space $\mathcal{C}_R(X)$ is called a *vector lattice*, if for each f, g in this subspace also the functions $f \vee g$ and $f \wedge g$, defined by $(f \vee g)(x) = \max(f(x), g(x))$ and $(f \wedge g)(x) = \min(f(x), g(x))$ at each point x in X, belong to this subspace. As for every f in $\mathcal{C}_R(X)$ we have $|f| = f \vee 0 + (-f) \vee 0$, we can reformulate the previous theorem as follows.

Theorem 3.53. *(Stone–Weierstrass Theorem, vector lattice version) If A is a separating vector lattice containing the identically 1 function in the space of all real valued continuous functions on the compact Hausdorff space X, then A is dense in $\mathcal{C}(X)$.*

M. H. Stone

T. W. Weierstrass

U. Dini

We shall need the following theorem, which is important in itself, too.

Theorem 3.54. *(Dini[15]) Given the functions $(f_n)_{n\in\mathbb{N}}$ and g in the space $\mathcal{C}_R(X)$ of all continuous, complex valued functions on the compact space X such that the sequence $(f_n(x))_{n\in\mathbb{N}}$ is decreasing and converges to $g(x)$ at each x in X, the function sequence $(f_n)_{n\in\mathbb{N}}$ converges to g in $\mathcal{C}_R(X)$.*

Proof. Given $\varepsilon > 0$ and x in X there exists a natural number n_x such that $f_{n_x}(x) - g(x) < \varepsilon$. By the continuity of f_{n_x} and g there is an open set U_x in X including x such that $f_{n_x}(y) - f_{n_x}(x) < \varepsilon$ and $g(x) - g(y) < \varepsilon$ holds for each y in U_x. As X is compact, there are elements $x_1, x_2, \dots, x_l$ in X (l is a positive integer) such that the sets $U_{x_1}, U_{x_2}, \dots, U_{x_l}$ cover X. For each $n \geqslant \max(n_{x_1}, n_{x_2}, \dots, n_{x_l})$ and for every x in U_{x_k} we have

$$f_n(x) - g(x) \leqslant f_{n_{x_k}}(x) - g(x)$$

$$= (f_{n_{x_k}}(x) - f_{n_{x_k}}(x_k)) + (f_{n_{x_k}}(x_k) - g(x_k)) + (g(x_k) - g(x)) < 3\varepsilon\,,$$

hence the function sequence $(f_n)_{n\in\mathbb{N}}$ converges to g in $\mathcal{C}_R(X)$. □

Theorem 3.55. *There is a sequence of real polynomials, which converges to the function $x \mapsto |x|$ in the normed space of all continuous real functions defined on the closed unit interval $[-1, 1]$.*

Proof. Let f_0 be the constant 1 polynomial. The real polynomials f_n defined for each real x and positive integer n by

$$f_n(x) = 2^{-1}(f_{n-1}^2(x) + (1 - x^2))$$

are nonnegative for $|x| \leqslant 1$, moreover $f_n(x) \geqslant f_{n+1}(x)$, as $f_{n-1}(x) \geqslant f_n(x)$ implies

$$f_n(x) - f_{n+1}(x)$$

$$= 2^{-1}(f_{n-1}^2(x) - f_n^2(x)) = 2^{-1}(f_{n-1}(x) + f_n(x))(f_{n-1}(x) - f_n(x)) \geqslant 0\,,$$

[15] Ulisse Dini, Italian mathematician (1845-1918)

and obviously $f_0(x) \geqslant f_1(x)$. It follows that the sequence $(f_n(x))_{n\in\mathbb{N}}$ converges to a real number $g(x) \leqslant 1$, for which we get, by taking limits in the formula defining f_n, the equation $g(x) = 2^{-1}(g^2(x) + (1 - x^2))$, that is $g(x) = 1 - |x|$. As g is continuous, hence, by Theorem 3.54, the sequence of polynomials $(1 - f_n)_{n\in\mathbb{N}}$ converges to the function $|x|$ in the normed space of continuous real functions defined on the closed interval $[-1, 1]$. □

The following theorem can be obtained by a simple linear transformation.

Theorem 3.56. *On every closed bounded interval the function $x \mapsto |x|$ is the uniform limit of a sequence of real polynomials.*

Now we are ready to prove the complex version of the Stone–Weierstrass Theorem.

Theorem 3.57. *(Stone–Weierstrass Theorem, complex version) If A is a closed separating subalgebra in the normed algebra $\mathcal{C}(X)$ of all complex valued continuous functions on the compact Hausdorff space X, which contains the constant 1 function, and also the complex conjugate of each element of it, then $A = \mathcal{C}(X)$.*

Proof. Let A_r denote the set of real valued functions in A. By Theorem 3.55, A_r includes the absolute values of the functions in A_r, too. If f is in A, then there are real functions u, v in $\mathcal{C}(X)$ such that $f = u + iv$, and here u and v necessarily belong to A, as the complex conjugate $\overline{f}$ of f satisfies $u = 2^{-1}(f + \overline{f})$ and $v = -2^{-1}i(f - \overline{f})$. Hence A_r is a separating family. By Theorem 3.52, A_r includes all real valued functions in $\mathcal{C}(X)$. It follows $A = \mathcal{C}(X)$. □

We recall that the *support* of the function $f : X \to \mathbb{C}$ is the closure of the set of all points at which the function is nonzero. This set is obviously closed, and we denote it by $\operatorname{supp} f$. The function f is called *compactly supported*, if its support is compact. The space of all continuous complex valued functions with compact support is denoted by $\mathcal{C}_c(X)$, on which the formula

$$\|f\| = \sup_{x\in X} |f(x)| \tag{3.9}$$

defines a norm. A linear functional of the space $\mathcal{C}_c(X)$ is called a *positive linear functional*, if it takes nonnegative real values on nonnegative real functions.

A more general class, including the class of compactly supported functions, is the class of functions *vanishing at infinity*. The function $f : X \to \mathbb{C}$ belongs this class, if for each $\varepsilon > 0$ there is a compact subset K in X such that the absolute value of f is smaller than ε on the complement of K. It is clear that formula (3.9) defines a norm on the linear space $\mathcal{C}_0(X)$ of all continuous functions of this type, moreover, this space turns to be a Banach space. Continuous functions vanishing at infinity on a locally compact Hausdorff space can be characterized easily. Let X be a locally compact Hausdorff space with its one-point compactification $\widetilde{X}$. For each $f : X \to \mathbb{C}$ let $\widetilde{f} : \widetilde{X} \to \mathbb{C}$ denote the unique function identical with f on X and taking 0 at ∞. Then we have the following theorem.

Theorem 3.58. *Let X be a locally compact Hausdorff space with one-point compactification $\widetilde{X}$. The mapping $f \longleftrightarrow \widetilde{f}$ sets up an isometric isomorphism between the Banach spaces $\mathcal{C}_0(X)$ and $\mathcal{C}(\widetilde{X})$.*

Proof. Clearly, $\widetilde{f}$ is continuous on $\widetilde{X}$ for each f in $\mathcal{C}_0(X)$, and the restriction of each continuous function on $\widetilde{X}$ to X is continuous on X. All other statements are obvious. □

Now we state a version of the Stone–Weierstrass Theorem on locally compact Hausdorff spaces.

Theorem 3.59. *(Stone–Weierstrass) If A is a closed separating subalgebra in the normed algebra $\mathcal{C}_0(X)$ of all complex valued continuous functions vanishing at infinity on the locally compact Hausdorff space X, which contains the conjugate of each element of it, further the functions in A do not have a common zero, then $A = \mathcal{C}_0(X)$.*

Proof. Let $\widetilde{X}$ denote the one-point compactification of X, for each f in $\mathcal{C}_0(X)$ let $\widetilde{f}$ denote the function defined above, finally let $\widetilde{A}$ be the set of all functions of the form $\widetilde{f}$ obtained from the functions f in A – together with the constant functions. Then $\widetilde{A}$ satisfies the conditions of the Stone–Weierstrass Theorem 3.57, hence it is identical with $\mathcal{C}(\widetilde{X})$, which implies our statement. □

Chapter 4

INVARIANT MEANS ON ABELIAN GROUPS

4.1 Means on Abelian groups

Invariant means on Abelian groups provide an efficient tool in diverse applications. They can be considered as substitutes of integral on all bounded functions. Here we need the existence of invariant means on arbitrary Abelian groups and their basic properties. For more about invariant means the reader should refer to [Greenleaf (1969)].

Let G be an Abelian group, and let $\mathcal{B}(G)$ denote the set of all bounded complex valued functions on G. Then $\mathcal{B}(G)$ is a complex Banach space with the norm

$$||f||_\infty = \sup_{x\in G} |f(x)| .$$

The translation operator τ_y corresponding to the element y in G is obviously a linear isometry on $\mathcal{B}(G)$.

A linear functional of $\mathcal{B}(G)$ is called a *mean* on G, if its value on nonnegative real functions is a nonnegative real number, and its value on the constant 1 function is 1. For instance, the evaluation functional at each point is a mean on G. As it is easy to see, every mean is a bounded linear functional, and all means form a nonempty convex set in the conjugate space $\mathcal{B}(G)^*$.

If M is a mean on the Abelian group G, and we have

$$M(\tau_y f) = M(f)$$

for each y in G and f in $\mathcal{B}(G)$, then M is called an *invariant mean* on G.

Theorem 4.1. *If G is a finite Abelian group, then the unique invariant mean on G is the functional I defined by*

$$I(f) = \frac{1}{|G|} \sum_{x \in G} f(x)$$

for each f in $\mathcal{B}(G)$.

Proof. Clearly, the above formula defines an invariant mean on G. Conversely, given an invariant mean M on G let ξ_x denote characteristic function of the one point set $\{x\}$, then, by the identity $\tau_{y^{-1}}(\xi_x) = \xi_{xy}$, it follows $M(\xi_x) = M(\xi_y)$ for each x, y in G. As $1 = \sum_{x \in G} \xi_x$ and $f = \sum_{x \in G} f(x)\xi_x$ holds for each f in $\mathcal{B}(G)$, we infer $M = I$. □

Theorem 4.2. *If M is a mean on the Abelian group G, then*

1. *$M(f) \leqslant M(g)$, whenever $f \leqslant g$;*
2. *$M(\overline{f}) = \overline{M(f)}$;*
3. *$|M(f)| \leqslant M(|f|)$;*
4. *$|M(f)| \leqslant ||f||$ holds for every f, g in $\mathcal{B}(G)$.*

Proof. If $f \leqslant g$, then $g - f$ is nonnegative, which implies 1. If $f = a + ib$, where a, b are real valued functions, then

$$M(\overline{f}) = M(a - ib) = M(a) - iM(b) ,$$

and

$$\overline{M(f)} = M(a) - iM(b)$$

follows, that is, we have 2. As $-|f| \leqslant f \leqslant |f|$, hence, by 1, it follows 3. Finally, 4 is a consequence of 3, as the function $\|f\| - |f|$ is real valued and nonnegative. □

Theorem 4.3. *Let G be an Abelian group. Then the set of all means on G is a weak*-compact set in $\mathcal{B}(G)^*$.*

Proof. The space $\mathcal{B}(G)^*$, endowed with the weak*-topology, is a locally convex topological vector space, in which the unit ball is compact, by the Banach–Alaoglu Theorem 3.42. As the set of all means on G is a subset of this set, to prove its compactness it is enough to show that it is a weak*-closed set. Let N be a bounded linear functional from the weak*-closure of the set of all means. Suppose that for some real valued function f we have $\operatorname{Im} N(f) \neq 0$. Then the set

$$\{\psi : |\psi(f) - N(f)| < |\operatorname{Im} N(f)|\}$$

is a weak*-neighborhood of N, hence it contains a mean φ, but this is impossible, as $\varphi(f)$ is a real number. Consequently, N takes real values on real functions. Suppose now that for some nonnegative real function f we have $N(f) < 0$. Then the set

$$\{\psi : |\psi(f) - N(f)| < -N(f)\}$$

is a weak*-neighborhood of N, hence it contains a mean φ, but this is impossible, as $\varphi(f)$ is nonnegative. It follows that N takes nonnegative real values on nonnegative real functions. Finally, for arbitrary $\varepsilon > 0$ the set $\{\psi : |\psi(1) - N(1)| < \varepsilon\}$ is a weak*-neighborhood of N, hence it contains a mean φ, but $\varphi(1) = 1$, hence $|1 - N(1)| < \varepsilon$, which implies $N(1) = 1$, that is, N is a mean. This proves that the set of all means on G is a weak*-compact set in $\mathcal{B}(G)^*$. □

4.2 Invariant means

In this section we present some basic properties of Abelian groups having invariant means. We call an Abelian group *amenable*, if there exists an invariant mean on it. In the previous section we proved the next theorem.

Theorem 4.4. *Every finite Abelian group is amenable.*

The following result is straightforward.

Theorem 4.5. *Every subgroup of an amenable group is amenable.*

Proof. Let M be an invariant mean on the amenable group G, and let H be a subgroup of G. We denote by Φ be the natural homomorphism of G onto the factor group with respect to H. For each f in $\mathcal{B}(H)$ the function $f \circ \Phi$ is in $\mathcal{B}(G)$, and we define $M^H(f) = M(f \circ \Phi)$. Then M^H is a mean on H. For any z in G/H let u be in $\Phi^{-1}(z)$, then, by

$$M^H(\tau_z f) = M\big(\tau_z f \circ \Phi\big) = M_x\big(f(\Phi(x) - z)\big)$$
$$= M_x\big(f(\Phi(x) - \Phi(u))\big) = M_x\big(f(\Phi(x - u))\big) = M(\tau_u(f \circ \Phi))$$
$$M(f \circ \Phi) = M^H(f)\,,$$

M^H is invariant. □

We used here the notation M_x to indicate that M is applied to the function in the bracket as a function of x.

Theorem 4.6. *Every homomorphic image of an amenable group is amenable.*

Proof. Let G be an amenable group, $\Phi : G \to H$ is a surjective homomorphism, and M an invariant mean on G. For each f in $\mathcal{B}(H)$ the function $f \circ \Phi$ is in $\mathcal{B}(G)$, and we let

$$M^H(f) = M(f \circ \Phi)\,.$$

It is straightforward that M^H is an invariant mean on H. □

Theorem 4.7. *Let G be an Abelian group and H be a subgroup. If H and G/H are amenable, then G is amenable.*

Proof. Let M_1 and M_2 be invariant means on H, and on G/H, respectively. For each function f in $\mathcal{B}(G)$ the function $x \mapsto M_1(\tau_x f)$ is constant on the cosets of H. Indeed, if $x - y$ lies in H, then $x = y + h$ with some h in H, and we have

$$\tau_x f = \tau_{y+h} f = \tau_h(\tau_y f)\,,$$

hence

$$M_1\big(\tau_h(\tau_x f)\big) = M_1(\tau_x f)\,.$$

Let the function $F_f : G/H \to \mathbb{C}$ be defined for $x + H$ in G/H by

$$F_f(x + H) = M_1(\tau_x f)\,,$$

then F_f is well-defined and bounded on G/H. Now we define

$$M(f) = M_2(F_f)\,,$$

then it follows immediately that M is an invariant mean on G. □

In particular, we have the following corollary.

Corollary 4.2.1. *The direct sum of two amenable groups is amenable.*

Theorem 4.8. *Let the Abelian group G be the union of the directed family $(H_\alpha)_{\alpha\in I}$ of its amenable subgroups: for each α, β in I there is a γ in I with $H_\alpha \cup H_\beta \subseteq H_\gamma$. Then G is amenable.*

Proof. Let M_α be an invariant mean on the subgroup H_α, then for each α in I the functional $M_\alpha^*(f) = M_\alpha(f|_{H_\alpha})$ is a mean on G, which is invariant with respect to translations with elements in H_α, were $f|_{H_\alpha}$ denotes the restriction of f to H_α. The set Λ_α of all means on G, which are invariant with respect to translations with elements in H_α, is a nonempty weak*-compact set in $\mathcal{B}(G)*$, and the family $(\Lambda_\alpha)_{\alpha\in I}$ is clearly centered. It follows that the intersection of these sets is nonempty, and it obviously consists of invariant means. □

In order to prove the main theorem of this section we need some additional tools. Namely, our proof depends on the Markov[1]–Kakutani[2] Fixed Point Theorem.

A. A. Markov

Sh. Kakutani

The mapping $T : C \to C$ on the convex subset $C \subseteq V$ of the complex linear space V is called *affine mapping*, if for each x, y in C and λ in $[0, 1]$ we have

$$T\big(\lambda x + (1-\lambda)y\big) = \lambda T(x) + (1-\lambda)T(y) .$$

Obviously, every linear mapping is affine.

Theorem 4.9. *(Markov–Kakutani Fixed Point Theorem) Every commuting family of continuous affine mappings of a nonempty, convex, compact subset in a locally convex topological vector space has a common fixed point.*

Proof. Let T be a mapping in the family, and for each x in the compact convex set C and $N = 0, 1, \ldots$ we define x_N by

$$x_N = \frac{1}{N+1} \sum_{n=0}^{N} T^n(x) .$$

As C compact, the sequence $(x_N)_{N\in\mathbb{N}}$ has a convergent subsequence $(x_{N_i})_{i\in\mathbb{N}}$ in C. Let y denote its limit. Let φ be an arbitrary continuous linear functional of the space. As C is compact, hence φ is bounded on C, that is $|\varphi(x)| \leqslant M$ holds for each x in C with some real number M. On the other hand, we have

$$|\varphi\big(T(x_N)\big) - \varphi(x_N)| = \frac{1}{N+1} |\varphi\big(T^{N+1}(x)\big) - \varphi(x)| \leqslant \frac{2M}{N+1} .$$

Let $N = N_i$, and let i tend to infinity, then $\varphi\big(T(y)\big) = \varphi(y)$ follows. As the conjugate space of a locally convex space is a separating family, we infer

[1]Andrei Andreyevich Markov, Russian mathematician (1856-1922)
[2]Shizuo Kakutani, Japanese mathematician (1911-2004)

$T(y) = y$, that is, each mapping in the family has a fixed point in C. For each mapping T in the family let C_T denote the set of all fixed points of T in C. As the mappings in the family commute, they all map C_T into itself, which gives that every finite subfamily of the family has a common fixed point in C_T. Clearly, the sets C_T are nonempty, convex, and compact further, they form a centered family. It follows that their intersection is nonempty, and it obviously consists of common fixed points of the given mappings. □

The main theorem follows.

Theorem 4.10. *Every Abelian group is amenable.*

Proof. If y is an element of the Abelian group G, and φ is a bounded linear functional on $\mathcal{B}(G)$, then we define the functional $T_y\varphi$ by the formula

$$T_y\varphi(f) = \varphi(\tau_y f)\,.$$

It is easy to see that $T_y\varphi$ is a bounded linear functional on $\mathcal{B}(G)$, and T_y is a bounded linear operator on the conjugate space $\mathcal{B}(G)^*$. Now we have the commuting family of continuous affine mappings $(T_y)_{y\in G}$ on the nonempty convex, compact set of all means on G, and, by Theorem 4.9, this family has a fixed point, which is obviously an invariant mean. □

Chapter 5

DUALITY OF DISCRETE AND COMPACT ABELIAN GROUPS

5.1 Characters on discrete Abelian groups

The homomorphisms of the discrete Abelian group G into the multiplicative group of the complex unit circle are called *characters*, which form – as it is easy to see – an Abelian group with respect to the pointwise multiplication of functions. This is called the *dual* of G. According to our previous notation we shall denote it by $\widehat{G}$, as in the case of finite Abelian groups this concept coincides with the one given above. We introduce a topology on the dual of G in the following way: let $\varepsilon > 0$ be arbitrary, and let K be a finite subset of G, further we define

$$U(K, \varepsilon) = \{\chi : |\chi(x) - 1| < \varepsilon \text{ for } x \in K\}.$$

It is easy to see that $\widehat{G}$, equipped with the topology, in which all sets $U(K, \varepsilon)$ form a neighborhood basis of the identity of the dual of G, is a topological group. In what follows when speaking about the dual of a discrete Abelian group as a topological group we always suppose that it is equipped with this topology.

Theorem 5.1. *The dual of a discrete Abelian group is a compact Abelian group.*

Proof. It is clear that in the dual of G a neighborhood basis of the identity is formed by the sets of the type

$$U(\{x\}, \varepsilon) = \{\chi : |\chi(x) - 1| < \varepsilon\},$$

where x is arbitrary in G. However, these sets form a neighborhood basis of the identity in the topology of the product space $\mathbb{T}^G$. The latter being compact, by Tikhonov's Theorem 3.4, it is enough to show that the dual of G is a closed set in this product topology. Let x, y be arbitrary points in G, and let $\varepsilon > 0$ a real number. Let μ denote an element in the closure of the dual of G in the product topology. As the set

$$\{\psi : |\psi(x) - \mu(x)| < \varepsilon\}$$

$$\cap\{\psi : |\psi(y) - \mu(y)| < \varepsilon\} \cap \{\psi : |\psi(xy) - \mu(xy)| < \varepsilon\}$$

is a neighborhood of μ, hence it includes an element χ of the dual of G. As $\chi(xy) = \chi(x)\chi(y)$, hence $|\mu(xy) - \mu(x) - \mu(y)| < 3\varepsilon$ holds. As $\varepsilon > 0$ is arbitrary, this implies that μ is a character. We infer that $\widehat{G}$ is a compact Abelian group. □

5.2 Characters on compact Abelian groups

The continuous homomorphisms of the compact Abelian group G into the multiplicative group of the complex unit circle are called *characters*, which form – as it is easy to see – an Abelian group with respect to the pointwise multiplication of functions. This is called the *dual* of G. According to our previous notation we shall denote it by $\widehat{G}$. We introduce a topology on the dual of G in the following way: let $\varepsilon > 0$ be arbitrary, and let K be a compact subset of G, further we define

$$U(K, \varepsilon) = \{\chi : |\chi(x) - 1| < \varepsilon \text{ for } x \in K\}.$$

It is easy to see that $\widehat{G}$, equipped with the topology, in which all sets $U(K, \varepsilon)$ form a neighborhood basis of the identity of the dual of G, is a topological group. In what follows when speaking about the dual of a compact Abelian group as a topological group we always suppose that it is equipped with this topology.

Theorem 5.2. *The dual of a compact Abelian group is a discrete Abelian group.*

Proof. In the dual of the compact Abelian group G the set

$$\{\chi : |\chi(x) - 1| < 1, \text{ if } x \in G\} \tag{5.1}$$

is a neighborhood of the identity. If χ is an element of it, then $\chi(G)$ is a compact subgroup of $\mathbb{T}$, different from $\mathbb{T}$. As $\mathbb{T}$ has no proper closed subgroup different from $\{1\}$, hence the above neighborhood is the singleton $\{1\}$, which means that the dual of G is discrete. □

5.3 Duality of discrete Abelian groups

Let G be a discrete or compact Abelian group. For each element x in G we define the function Φ_x on the dual group of G as follows: for every χ in $\widehat{G}$ we let

$$\Phi_x(\chi) = \chi(x) .$$

As it is easy to see, Φ_x is a character of $\widehat{G}$, that is, the mapping $\Phi : x \mapsto \Phi_x$ maps G into its second dual. This mapping is called the *natural mapping* of G. We should call it *natural homomorphism* in the light of a forthcoming theorem.

An Abelian group is called *divisible*, if the mapping $x \mapsto n \cdot x$ is surjective for each positive integer n. We shall need the following theorem (see Theorem (A.7) in [Hewitt and Ross (1979)], p. 441).

Theorem 5.3. *Every homomorphism of a subgroup in an Abelian group into a divisible Abelian group can be extended to a homomorphism of the whole group.*

Proof. Let H be a subgroup of the discrete Abelian group G, and φ a homomorphism of H into the divisible Abelian group D. The set of all extensions of φ to the subgroups including H form a partially ordered set with respect to inclusion. It is easy to see that in this partially ordered set every nonempty chain has an upper bound, hence, by Zorn's Lemma 3.2, this set has a maximal element, which is obviously a maximal homomorphic extension of φ. We show that its domain is the whole group G. To do this it is enough to show that if an element X does not belong to H, then φ has a homomorphic extension to the subgroup H_0 of all elements of the form $x^n h$, where n is an integer and h is in H. We consider two cases. If x^n does not belong to H for all $n \geqslant 2$ integers, then we let

$$\tilde{\varphi}(x^n h) = \varphi(h)$$

whenever $x^n h$ is in H_0. Then $\tilde{\varphi}$ is well-defined, because if $x^{n_1} h_1 = x^{n_2} h_2$ holds for some h_1, h_2 in H, and n_1, n_2 integers, then $x^{n_1 - n_2}$ lies in H, which is impossible. In other words, in this case the representation of the elements of H_0 in the form $x^n h$ with an integer n and h in H is unique. Hence

$$\begin{aligned}\tilde{\varphi}(x^{n_1} h_1 \cdot x^{n_2} h_2) &= \tilde{\varphi}(x^{n_1} x^{n_2} \cdot h_1 h_2) = \varphi(h_1)\varphi(h_2)\\ &= \tilde{\varphi}(x^{n_1} h_1) \cdot \tilde{\varphi}(x^{n_2} h_2) ,\end{aligned}$$

that is, $\tilde{\varphi}$ is a homomorphism of G into D, which clearly agrees with φ on H.

In the opposite case there is an integer $n \geqslant 2$ such that x^n belongs to H. Let k be the smallest integer with this property. We choose an element z in D with $z^k = \varphi(x^k)$, and we define

$$\tilde{\varphi}(x^n h) = z^n \varphi(h).$$

We have to show that the right hand side is independent of the particular choice of n and h, that is, if $x^{n_1} h_1 = x^{n_2} h_2$, then

$$z^{n_1} \varphi(h_1) = z^{n_2} \varphi(h_2).$$

There exist natural numbers q, l such that $0 \leqslant l < k$ and $n_2 - n_1 = qk + l$. As

$$x^{n_2 - n_1} = x^{qk+l} = (x^k)^q \cdot x^l,$$

further $(x^k)^q$ is in H, so does x^l, and, by the minimality of k, it follows $l = 0$. We infer

$$z^{n_2} \varphi(h_2) = z^{n_1} z^{qk} \varphi(h_2) = z^{n_1} \varphi(x^{qk} h_2),$$

but, by the equality $x^{n_1} h_1 = x^{n_2} h_2 = x^{n_1} x^{qk} h_2$, we have $x^{qk} h_2 = h_1$, hence the right hand side of the previous equation is $z^{n_1} h_1$.

Finally, we have

$$\begin{aligned} \tilde{\varphi}(x^{n_1} h_1 \cdot x^{n_2} h_2) &= \tilde{\varphi}(x^{n_1+n_2} \cdot h_1 h_2) = z^{n_1+n_2} \varphi(h_1) \varphi(h_2) \\ &= \tilde{\varphi}(x^{n_1} h_1) \cdot \tilde{\varphi}(x^{n_2} h_2), \end{aligned}$$

hence $\tilde{\varphi}$ is a homomorphism of G into D, which agrees with φ on H. The theorem is proved. □

Theorem 5.4. *Discrete Abelian group is isomorphic to its second dual via the natural homomorphism.*

Proof. The natural mapping is a homomorphism of G into its second dual, which is shown by the equality

$$\Phi_{xy}(\chi) = \chi(xy) = \chi(x)\chi(y) = \Phi_x(\chi)\Phi_y(\chi).$$

The kernel of Φ consists of those elements of G satisfying $\chi(x) = \Phi_x(\chi) = 1$ for each character χ. We show that this is possible only if $x = e$. Indeed, let $x \neq e$ be arbitrary in G, and we choose an $\alpha \neq 1$ complex number with modulus 1. The mapping $x^k \mapsto \alpha^k$, where k is an integer, is clearly a homomorphism of the subgroup generated by x into $\mathbb{T}$. As $\mathbb{T}$ is divisible, by the previous theorem, this homomorphism extends to a character of G, and the extension takes the value α at x, which is different from 1. This proves that the natural homomorphism is injective.

Now we prove surjectivity. The dual of G is a compact Abelian group, and $\Phi(G)$ is a separating family in $\mathcal{C}(\widehat{G})$. Indeed, if $\chi_1 \neq \chi_2$ are in $\widehat{G}$, then $\chi_1(x) \neq \chi_2(x)$ holds for some x in G, that is, we have $\Phi_x(\chi_1) \neq \Phi_x(\chi_2)$. We consider the set of all finite linear combinations of the functions in $\Phi(G)$. These form a separating algebra in $\mathcal{C}(\widehat{G})$, which includes the function 1 and is closed with respect to complex conjugation. Then, by the Stone–Weierstrass Theorem 3.57, $\Phi(G)$ is dense in $\mathcal{C}(\widehat{G})$. If ξ is a character of $\widehat{G}$, which does not belong to $\Phi(G)$, then ξ is orthogonal to each element of $\Phi(G)$, as different characters are orthogonal. Hence it is orthogonal to $\mathcal{C}(\widehat{G})$. But $\mathcal{C}(\widehat{G})$ is dense in $L^2(\widehat{G})$, hence ξ is orthogonal to $L^2(\widehat{G})$, including ξ itself, a contradiction. The proof is complete. □

5.4 Convolution operators on compact Abelian groups

Let I be a fixed invariant mean on the compact Abelian group G. As every continuous function on G is bounded, hence I is a bounded linear functional on $\mathcal{C}(G)$, having all properties listed in Theorem 4.2, and $I(1) = 1$.

Theorem 5.5. *If f is not identically zero, nonnegative, and continuous real function on the compact Abelian group, then $I(f) > 0$.*

Proof. As f is not identically zero, there is a point x_0, an open neighborhood U of the identity, and a positive number $m > 0$ such that $f(x) \geqslant m$, whenever x is in $x_0 U$. All sets of the form yU with y in G form an open covering G, hence, by compactness, there are elements $y_1, y_2, \ldots, y_n$ in G such that

$$G \subseteq y_1 U \cup y_2 U \cup \cdots \cup y_n U \,.$$

Then we have

$$\tau_{y_1^{-1}} f + \tau_{y_2^{-1}} f \cdots + \tau_{y_n^{-1}} f \geqslant m \,,$$

which implies

$$I(f) = \frac{1}{n} \sum_{j=1}^{n} I(\tau_{a_j^{-1}} f) = \frac{1}{n} I(\sum_{j=1}^{n} \tau_{a_j^{-1}} f) \geqslant \frac{m}{n} > 0 \,.$$

□

As it is easy to see, the formula

$$\langle f, g \rangle = I(f \cdot \overline{g})$$

defines an inner product on $\mathcal{C}(G)$. Then $\mathcal{C}(G)$, equipped with the corresponding norm

$$\|f\|_2 = (I(|f|^2))^{\frac{1}{2}}$$

is a normed space, however, it is not necessarily complete. Its completion is denoted by $L^2(G)$ to which the inner product extends in a unique manner, hence it becomes a Hilbert space. If f is in $L^2(G)$, then $I(f)$ means the inner product $\langle f, 1\rangle$. This obviously agrees with the original value of $I(f)$, if f is continuous. The functional I, which is now defined on $\mathcal{C}(G)$, and is obviously a bounded linear functional, is called *Haar*[1] *integral.*

It seems reasonable to repeat the argument applied in the case of finite Abelian groups with the Hilbert space $L^2(G)$ instead of $\mathcal{C}(G)$. The translation operators are again commuting unitary operators, having the characters as common eigenfunctions. However, at this moment there is no guarantee that the translation operators do have sufficiently enough common eigenfunctions in $L^2(G)$, hence the question arises: does there exist a complete orthonormal set consisting of common eigenfunctions of the translation operators? Instead of translation operators we shall rather consider convolution operators.

A. L. Cauchy

V. J. Bunyakovszkij

H. A. Schwarz

If f, g are in $L^2(G)$, then, by the well-known Cauchy[2]–Bunyakovszkij[3]–
–Schwarz[4] inequality the function $y \mapsto f(xy^{-1})g(y)$ is in $L^2(G)$, and the *convolution* is defined by

$$(f * g)(x) = I_y(f(xy^{-1})g(y)) \tag{5.2}$$

for each x in G. We use the notation I_y to indicate that the functional I is applied to the function $y \mapsto f(xy^{-1})g(y)$ with x is fixed. Then, as it

[1]Alfréd Haar, Hungarian mathematician (1885-1933)
[2]Augustin Louis Cauchy, French mathematician (1789-1857)
[3]Viktor Yakovlevich Bunyakovsky, Russian mathematician (1804-1889)
[4]Hermann Amandus Schwarz, German mathematician (1843-1921)

is easy to see, the function $f * g$ belongs to $L^2(G)$, too, moreover, if f is continuous, then so is $f * g$.

For each function a in $\mathcal{C}(G)$ the operator $A_a : f \mapsto a * f$ is obviously linear, and it is called the *convolution operator* corresponding to a. It is easy to check that A_a is a *compact operator*: the closure of the image of each bounded set is compact.

The following theorem is fundamental in the theory of compact operators: the Spectral Theorem of compact normal operators (see e.g. [Arveson (2002)], Theorem 2.8.2, p. 69). The following result is a particular case of it.

Theorem 5.6. *(Hilbert–Schmidt[5] Theorem) Every nonzero compact normal operator on a Hilbert space has a nonzero eigenvalue, and every eigenspace corresponding to nonzero eigenvalues is finite dimensional.*

In our case, it is easy to see that $A_a^* = A_{a^*}$, where $a^*(x) = \overline{a(x^{-1})}$. Hence every convolution operator is normal. Which functions are the common eigenfunctions of all convolution operators? If the function f in $L^2(G)$ is one of them, then for each a in $L^2(G)$ we have $a * f = \lambda(a) \cdot f$. It is also easy to see that the intersection of the kernels of the convolution operators is 0. Indeed, if $a * f = 0$ holds for each a in $L^2(G)$, with $a = f^*$ we have the following equality

E. Schmidt

$$0 = f^* * f(e) = \int_G f^*(y^{-1}) f(y) dy = \int |f(y)|^2 dy \,.$$

It follows that for each common eigenfunction f there is an a with $\lambda(a) \neq 0$. On the other hand

$$\tau_z(a * f)(x) = \int_G a(xzy^{-1}) f(y) dy = \tau_z a * f(x) = \lambda(\tau_z a) \cdot f(x)$$

holds for each z in G, hence

$$\tau_z f = \frac{1}{\lambda(a)} \tau_z(a * f) = \frac{\lambda(\tau_z a)}{\lambda(a)} \cdot f \,,$$

[5]Erhard Schmidt, German mathematician (1876-1959)

which means that f is also a common eigenfunction of all translation operators. Conversely, if f is a common eigenfunction of all translation operators and $f(e) = 1$, then f is a character, and it is a common eigenfunction of all convolution operators. The following fundamental result implies that in $L^2(G)$ there is a complete orthonormal set consisting of all common eigenfunctions of the convolution operators (see e.g. [Arveson (2002)], Theorem 2.8.2, p. 69).

Theorem 5.7. *(Hilbert–Schmidt Theorem) Given a nonempty family of commuting compact normal operators on a Hilbert space, then there exists a complete orthonormal set consisting of all common eigenfunctions of these operators.*

Proof. Let $\mathcal{A}$ be the family of the operators with the given properties. If $\mathcal{A} = \{0\}$, then the statement is trivial. If $A \neq 0$ is in $\mathcal{A}$, then it has a nonzero eigenvalue, and the corresponding eigenspace is finite dimensional. Due to the commuting property of the family this eigenspace is invariant under all operators in $\mathcal{A}$. In this subspace all operators in $\mathcal{A}$, different from A, have a common eigenvector, which is obviously an eigenvector of A, too. Hence there exists a common eigenvector of all operators in $\mathcal{A}$. Each operator in $\mathcal{A}$ can be written in the form $A = A_1 + iA_2$, where A_1, A_2 are self adjoint, compact, and commuting. The set of all operators arising in this way we denote by $\mathcal{B}$. If A is a normal and B is an arbitrary operator with $AB = BA$, then $AB^* = B^*A$. Hence $\mathcal{B}$ consists of compact self adjoint commuting operators. The common eigenvectors of $\mathcal{B}$ and $\mathcal{A}$ are the same. Let a maximal orthonormal set of common eigenvectors of the operators in $\mathcal{B}$ be given, and let S denote the closure of its linear span. If the orthogonal complement of S is different from $\{0\}$, then this orthogonal complement is invariant under the operators in $\mathcal{B}$, hence there is a common eigenvector in it, contradicting to maximality. This means that the maximal orthonormal system is complete, and the theorem is proved. □

By this theorem, given a compact Abelian group G there exists a complete orthonormal set in $L^2(G)$ consisting of characters. As different characters are orthogonal, all characters must be included in this complete orthonormal set.

Theorem 5.8. *If G is a compact Abelian group, then all characters of G form a complete orthonormal set in $L^2(G)$.*

Now we have a simple proof for the uniqueness of the Haar integral on compact Abelian groups.

Theorem 5.9. *On compact Abelian groups the Haar integral is unique.*

Proof. Suppose that I_1 and I_2 are two Haar integrals, and the corresponding L^2 spaces are $L_1^2(G)$ and $L_2^2(G)$. On finite linear combinations of characters the two inner products are the same, and, as these are dense in $L_1^2(G)$ and in $L_2^2(G)$, it follows that the inner product agrees on $L_1^2(G)$ and on $L_2^2(G)$, hence so does it on their intersection, that is, on $\mathcal{C}(G)$. □

5.5 Duality of compact Abelian groups

We shortly summarize our knowledge about the duals of discrete and compact Abelian groups. Let G be a discrete Abelian group, and let $\widehat{G}$ be the set of all characters of G with the topology inherited from $\mathbb{C}^G$. By Tikhonov's Theorem 3.4, the set

$$\{z \in \mathbb{C} : |z| \leqslant 1\}^G$$

is compact, and $\widehat{G}$ is a closed subspace of it, hence $\widehat{G}$ is compact, too. All sets of the type

$$\{\chi \in \widehat{G} : |\chi(g_k) - 1| < \varepsilon, k = 1, 2, \ldots, n\}$$

form a neighborhood basis of the identity, where $\varepsilon > 0$ is an arbitrary positive number, $n \geqslant 1$ is a natural number, and $g_1, g_2, \ldots, g_n$ a G are arbitrary elements in G. Then $\widehat{G}$ is a compact Abelian group.

Let G be a compact Abelian group, and let $\widehat{G}$ be the set of all continuous characters of G with the topology in which a neighborhood basis of the identity is formed by all sets of the type

$$\{\chi \in \widehat{G} : |\chi(g) - 1| < \varepsilon, g \in K\},$$

where $\varepsilon > 0$ is an arbitrary positive number, and $K \subseteq G$ is an arbitrary compact set. As G is compact, hence $\{1\}$ is a neighborhood of the identity in $\widehat{G}$, which implies that $\widehat{G}$, equipped with this topology is a discrete Abelian group.

Both of these topologies are special cases of the following.

Let G be a locally compact Abelian group, and let $\widehat{G}$ be the set of all continuous characters of G, in which a neighborhood basis of the identity is formed by all sets of the type

$$\{\chi \in \widehat{G} : |\chi(g) - 1| < \varepsilon, g \in K\},$$

where $\varepsilon > 0$ is an arbitrary positive number, and $K \subseteq G$ is an arbitrary compact set. Equipped with this topology $\widehat{G}$ is a locally compact Abelian group. We call this topology *compact-open topology.*

Let G be a locally compact Abelian group, and let $\Phi : G \to \widehat{\widehat{G}}$ be the natural homomorphism of G into its second dual, that is

$$\Phi(g)(\chi) = \chi(g),$$

whenever g is in G and χ is a continuous character of G. The basic project in duality theory is to show that Φ is a topological isomorphism of G onto $\widehat{\widehat{G}}$. This is the famous Pontryagin Duality Theorem. For more about Pontryagin's Duality the reader should refer to [Loomis (1953); Reiter (1968); Dixmier (1969); Hewitt and Ross (1979); Rudin (1990)]. If G is discrete, then $\widehat{G}$ is compact, hence $\widehat{\widehat{G}}$ is discrete, which implies that Φ is continuous, and if it is surjective and invertible, then its inverse is also continuous. This means that for the duality of discrete Abelian groups it is necessary and sufficient that the natural homomorphism is injective and surjective, that is, it is an algebraic isomorphism of G onto its second dual. This is what we have seen in Section 5.3. Now we shall consider the case of compact Abelian groups.

Suppose that G is a compact Abelian group. Its dual $\widehat{G}$ is a discrete group, and $\widehat{\widehat{G}}$ is compact again. By the definition of the natural homomorphism $\Phi : G \to \widehat{\widehat{G}}$ is obviously continuous. We show that Φ is injective and surjective.

Theorem 5.10. *The natural homomorphism of a compact Abelian group into its second dual is a topological isomorphism.*

Proof. To prove injectivity it is enough to show that for each element g in G different from the identity there is a character χ of G for which $\chi(g) \neq 1$. Supposing the contrary the operator τ_g is the identity on $L^2(G)$, hence it is the identity on $\mathcal{C}(G)$, too: we have $f(xg) = f(x)$ for each continuous function f and for every x in G, but this is impossible, as $xg \neq x$, hence these two elements can be separated by a continuous function, by Theorems 3.10 and 3.13. By Theorem 3.15, it follows that the natural homomorphism is a topological isomorphism of G onto $\Phi(G)$. If $\Phi(G) \neq \widehat{\widehat{G}}$, then the factor group $\widehat{\widehat{G}}/\Phi(G)$ is a nontrivial compact Abelian group, which has a nontrivial character. By the composition of the homomorphisms $\widehat{\widehat{G}} \mapsto \widehat{\widehat{G}}/\Phi(G) \mapsto \mathbb{T}$ this character induces on $\widehat{\widehat{G}}$ a nontrivial character ξ satisfying $\xi(\Phi(G)) = \{1\}$. As $\widehat{G}$ is discrete, and we have proved the duality theorem for discrete groups, it follows that ξ is an element of $\widehat{G}$, that is, it is a nontrivial character of G, but $\xi(g) = 1$ holds for each g in G, a contradiction. The theorem is proved. □

Summarizing our results, we have that if G is a discrete or a compact Abelian group, then it is topologically isomorphic to its second dual, and the topological isomorphism is provided by the natural homomorphism.

Chapter 6

DUALITY OF ELEMENTARY ABELIAN GROUPS

6.1 The dual of some special Abelian groups

We shall need the following technical result.

Theorem 6.1. *The dual of a compact Abelian group has no proper separating subgroup.*

Proof. Suppose that Γ is a separating subgroup in the dual of the compact Abelian group G. Then for each g in G, different from the identity, there is a character χ in Γ such that $\chi(g) \neq 1$. The set of all finite linear combinations of the characters in Γ satisfies the conditions of the Stone–Weierstrass Theorem 3.57, hence they form a dense subset in $\mathcal{C}(\widehat{G})$. If there is a character χ of G not in Γ, then χ is orthogonal to Γ, hence it is orthogonal to $\mathcal{C}(\widehat{G})$, too, that is, to itself, which is impossible. $\square$

Theorem 6.2. $\widehat{\mathbb{T}} = \mathbb{Z}$.

Proof. The identity mapping $\chi(g) = g$ (g is in $\mathbb{T}$) is a separating character on $\mathbb{T}$, hence it generates $\widehat{\mathbb{T}}$. On the other hand, the mapping $\chi^n \mapsto n$ (n is an integer) is an isomorphism between $\widehat{\mathbb{T}}$ and $\mathbb{Z}$. $\square$

Theorem 6.3. $\widehat{\mathbb{Z}} = \mathbb{T}$.

Proof. The statement follows from the Duality Theorem 5.10 on compact Abelian groups, and from Theorem 6.2. $\square$

Theorem 6.4. *If $G_1, G_2, \ldots, G_n$ are locally compact Abelian groups, then $(G_1 \times G_2 \times \cdots \times G_n)\widehat{\ }$ is topologically isomorphic to $\widehat{G}_1 \times \widehat{G}_2 \times \cdots \times \widehat{G}_n$.*

Proof. We prove the statement for $n = 2$, the general case can be treated similarly.

Let e_1, resp. e_2 be the identity in G_1, resp G_2, and let for each χ_1 in $\widehat{G}_1$ and χ_2 in $\widehat{G}_2$

$$\Theta(\chi_1, \chi_2)(x_1, x_2) = \chi_1(x_1)\chi_2(x_2)\,,$$

whenever x_1 is in G_1, and x_2 is in G_2. Then $\Theta(\chi_1, \chi_2)$ is a character of $G_1 \times G_2$, and it is easy to see that Θ is a homomorphism of $\widehat{G}_1 \times \widehat{G}_2$ into the dual of $G_1 \times G_2$. If we let $\Theta(\chi_1, \chi_2)(x_1, x_2) = 1$ for each x_1 in G_1 and x_2 in G_2, then $\chi_1(x_1)\chi_2(x_2) = 1$ follows, hence, by the successive substitutions $x_1 = e_1$ and $x_2 = e_2$, we get $\chi_1 = \chi_2 = 1$, that is, Θ is an isomorphism. On the other hand, for each χ in $(G_1 \times G_2)\hat{}$ we have

$$\chi(x_1, x_2) = \chi(x_1, e_2)\chi(e_1, x_2)\,,$$

further $x \mapsto \chi(x, e_2)$ and $x \mapsto \chi(e_1, x)$ are characters of G_1 and G_2, hence χ is in the set $\Theta(\widehat{G}_1 \times \widehat{G}_2)$. Consequently, Θ is surjective, and $\widehat{G}_1 \times \widehat{G}_2$ is algebraically isomorphic to $(G_1 \times G_2)\hat{}$. Now we show that Θ is a topological isomorphism, that is, it is continuous and open.

Let F be a compact subset in $G_1 \times G_2$, and let $\varepsilon > 0$ be an arbitrary real number. If F_1 and F_2 denote the projections of F into G_1, resp. G_2, then F_1 and F_2 are compact sets, further

$$\Theta(U(F_1, \varepsilon/2) \times U(F_2, \varepsilon/2) \subseteq U(F, \varepsilon)\,,$$

hence Θ is continuous. On the other hand, if F_1 is compact in G_1, and F_2 is compact in G_2, further $\varepsilon_1 > 0$ and $\varepsilon_2 > 0$ are real numbers, then the set

$$F = (F_1 \cup \{e_1\}) \times (F_2 \cup \{e_2\})$$

is compact, and with the notation $\varepsilon = \min(\varepsilon_1, \varepsilon_2)$ we have

$$U(F, \varepsilon) \subseteq \Theta(U(F_1, \varepsilon_1) \times U(F_2, \varepsilon_2))\,,$$

which proves that Θ is open. □

Theorem 6.5. *If F is a finite Abelian group, then $\widehat{F} = F$.*

Proof. Every finite Abelian group is a direct product of cyclic groups:

$$F = \mathbb{Z}(m_1) \times \mathbb{Z}(m_2) \times \cdots \times \mathbb{Z}(m_k) \tag{6.1}$$

($k, m_1, m_2, \ldots, m_k$ are positive integers) holds, where $\mathbb{Z}(m) = \mathbb{Z}/m\mathbb{Z}$ is the cyclic group of order m. The group $\mathbb{Z}(m)$ is isomorphic to the multiplicative group of m-th roots of unity. On this group the identity mapping is a separating character, hence its powers fill $\mathbb{Z}(m)\hat{}$, but the group of its powers is isomorphic to $\mathbb{Z}(m)$ itself. Hence $\mathbb{Z}(m)\hat{} = \mathbb{Z}(m)$, and, by the previous theorem, our statement follows. □

Theorem 6.6. $\widehat{\mathbb{R}} = \mathbb{R}$.

Proof. Let f be a character of $\mathbb{R}$, then it can be written in the form $f(x) = \exp(it(x))$ (x is a real number), where t is a continuous real valued function. The identity

$$\exp(it(x+y)) = f(x+y) = f(x)f(y) = \exp(i[t(x)+t(y)])$$

(x, y are real numbers) implies $t(x+y) - t(x) - t(y) = n(x,y)$ (x, y are real numbers), where n is a continuous function on $\mathbb{R} \times \mathbb{R}$ having integer values. As $\mathbb{R} \times \mathbb{R}$ is connected, hence n is constant, and, by $n(0,0) = 0$, we infer $t(x+y) = t(x) + t(y)$ for every real x, y. This gives $t(x) = \lambda x$ (x is a real number) with some real λ, hence $f(x) = \exp(\lambda x)$ (x is a real number). It is easy to see that the mapping $f \to \lambda$ is a topological isomorphism between $\mathbb{R}$ and $\widehat{\mathbb{R}}$. □

6.2 Elementary Abelian groups

Theorem 6.7. *If F is a finite Abelian group, then*

$$(\mathbb{R}^n \times \mathbb{Z}^k \times \mathbb{T}^m \times F)\widehat{\ } = \mathbb{R}^n \times \mathbb{T}^k \times \mathbb{Z}^m \times F$$

(n, k, m are nonnegative integers).

Proof. Obvious. □

The groups, which are topologically isomorphic to groups of the form $\mathbb{R}^n \times \mathbb{Z}^k \times \mathbb{T}^m \times F$ (n, k, m are nonnegative integers), where F is a finite Abelian group, are called *elementary Abelian groups.*

Theorem 6.8. *The natural mapping on an elementary Abelian group is a topological isomorphism of the group onto its second dual.*

Proof. Obvious. □

The most important types of elementary Abelian groups are the following ones: the compact elementary Abelian groups are of the form $\mathbb{T}^m \times F$ (m is a natural number, F is a finite Abelian group); the discrete elementary Abelian groups are of the form $\mathbb{Z}^k \times F$ (k is a natural number, F is a finite Abelian group); the connected elementary Abelian groups are of the form $\mathbb{R}^n \times \mathbb{T}^m$ (n, m are natural numbers); the so-called *vector groups* are of the form $\mathbb{R}^n \times \mathbb{Z}^k$ (n, k are natural numbers), and the so-called *lattice groups* are of the form $\mathbb{Z}^k$ (k is a natural number).

Chapter 7

HARMONIC ANALYSIS ON COMPACT ABELIAN GROUPS

7.1 The Riesz Representation Theorem

We shall set up harmonic analysis on compact Abelian groups along the lines of what we have done on finite Abelian groups. In the finite case we used effectively the arithmetic mean of the function values, which served as a basic tool, and was replaced later, considering infinite discrete Abelian groups, by an invariant mean. On compact Abelian groups a similar role is played by the Haar integral, whose uniqueness was proved in Theorem 5.9. The terminology suggests the appearance of an integral, and the question arises immediately: from which measure originates this integral concept? The answer is provided by the famous Riesz[1] Representation Theorem. This theorem, together with its different versions, describes the conjugate space of various topological vector spaces setting up a connection between linear functionals and measures (see e.g. [Holmes (1975)], Chapter IV). For general measure theory see [Halmos (1950)]. First we need some auxiliary concepts.

Integration theory on locally compact Hausdorff topological spaces depends on the concept of Borel measures. We recall that a *measurable space* is an ordered pair consisting of a nonempty set and a *σ-algebra* of its subsets. Let X be a locally compact Hausdorff topological space, and let Ω denote the σ-algebra generated by the open sets. The sets belonging to Ω are called *Borel*[2] *sets.* By the Hausdorff property and Theorem 3.12, every compact set is closed, hence it is a Borel set. Measures defined on the measurable space (X, Ω) are called *Borel measures.*

[1]Frigyes Riesz, Hungarian mathematician (1880-1956)

[2]Félix Edouard Justin Émile Borel, French mathematician (1871-1956)

A Borel measure μ is called *regular* on X, if the measure of every compact set is finite, the measure of every Borel set of finite measure is equal to the supremum of the measures of all compact subsets of the set, and the measure of every Borel set is equal to the infimum of the measures of all open sets containing the set. A complex measure is called a Borel measure, if its absolute value is a Borel measure.

E. Borel

Using the Hahn[3]–Jordan[4] Decomposition of complex measures (see e.g. [Halmos (1950)], § 29., p. 120) it is easy to see, that each complex measure is a complex linear combination of four nonnegative measures. It is also easy to check that a complex measure is a Borel measure if and only if all these four measures can be taken as Borel measures. Clearly, all regular complex Borel measures on X form a complex linear space, which will be denoted by $\mathcal{M}(X)$, and the elements of it will simply be called *measures.* The *support* of a measure μ is the complement of the union of all open sets having μ measure zero, which is then a closed set. We denote it by $\operatorname{supp} \mu$. If it is compact, then the measure μ is said to be *compactly supported.* All compactly supported measures form a subspace in $\mathcal{M}(X)$, which we denote by $\mathcal{M}_c(X)$. As it is easy to see, the formula

$$\|\mu\| = |\mu|(X) \tag{7.1}$$

defines a norm on the space $\mathcal{M}(X)$.

H. Hahn

C. Jordan

After this introduction we proceed with the main theorem.

[3]Hans Hahn, Austrian mathematician (1879-1934)

[4]Marie Ennemond Camille Jordan, French mathematician (1838-1922)

Theorem 7.1. *(Riesz Representation Theorem) Let X be a locally compact Hausdorff space, and for each complex Borel measure μ we define*

$$F_\mu(f) = \int f \, d\mu \,, \tag{7.2}$$

whenever f is in $\mathcal{C}_0(X)$. Then $F_\mu(f)$ is a bounded linear functional on $\mathcal{C}_0(X)$, and the mapping $\mu \mapsto F_\mu$ is an isometric isomorphism of $\mathcal{M}(X)$ onto $\mathcal{C}_0(X)^*$.

Hence this theorem characterizes the conjugate space of the Banach space of all continuous complex valued functions vanishing at infinity on a locally compact Hausdorff space by describing its bounded linear functionals: these are exactly integrals with respect to Borel measures.

Another version of this theorem follows.

Theorem 7.2. *(Riesz Representation Theorem) Let X be a locally compact Hausdorff space, and let F be a positive linear functional on $\mathcal{C}_c(X)$. Then there exists a unique regular Borel measure μ on X such that*

$$F(f) = \int f \, d\mu \,, \tag{7.3}$$

holds, whenever f is in $\mathcal{C}_c(X)$.

The corresponding result for compact Hausdorff spaces is the following theorem. In this case $\mathcal{C}(X)$ obviously consists of bounded functions, and the norm (3.9) induces a Banach space structure on it.

Theorem 7.3. *(Riesz Representation Theorem) If X is a compact Hausdorff space, then the mapping $\mu \mapsto F_\mu$, where F_μ is defined by (7.2) for each function f in $\mathcal{C}(X)$, is an isometric isomorphism of $\mathcal{M}(X)$ onto $\mathcal{C}(X)^*$.*

We have seen in Section 5.4 that the Haar integral I, defined uniquely on the compact Abelian group G, is a linear functional of the space $\mathcal{C}(G)$, to which there corresponds, by the previous theorem, a unique regular complex Borel measure such that

$$I(f) = \int f \, d\mu$$

holds for each continuous function $f : G \to \mathbb{C}$. By Theorem 4.2, the properties of I imply that μ is a measure. The measure μ is called *Haar measure*.

Alfréd Haar

Frigyes Riesz

In the following theorem we summarize the most important properties of the Haar measure on compact Abelian groups.

Theorem 7.4. *Let G be a compact Abelian group. The Haar measure is the unique positive measure on the family of Borel sets having the following properties:*

1. $\mu(G) = 1$.
2. *The measure μ is translation invariant: for each Borel set B and g in G we have $\mu(g \cdot B) = \mu(B)$.*
3. *The measure of each nonempty open set is positive.*
4. *The measure of each open set is equal to the supremum of the measures of all compact subsets of it.*
5. *The measure of each Borel set is equal to the infimum of the measures of all open sets containing it.*

The first three statements follow from the properties of the invariant measure, while the last two statements and the uniqueness are consequences of the Riesz Representation Theorem 7.3.

7.2 Haar measure on the complex unit circle

It is clear that every finite group is compact, if equipped with the discrete topology. As it is easy to see, this is the only Hausdorff topology, which is compatible with the group operations, in other words, this is the only Hausdorff topology, which makes a finite group a topological group. On finite groups the counting measure divided by the number of the elements in the group obviously satisfies the conditions listed in the previous theorem, hence, by uniqueness, it is identical to the Haar measure. The next simplest compact Abelian group is $\mathbb{T}$, the complex unit circle. The following theorem refers to this group.

Theorem 7.5. *Let $f : \mathbb{T} \to \mathbb{C}$ be a continuous function, and let μ be the Haar measure on $\mathbb{T}$. Then we have*

$$\int f \, d\mu = \frac{1}{2\pi} \int_0^{2\pi} f(e^{it}) \, dt \, .$$

In particular, if B is a Borel set in $\mathbb{T}$, then

$$\mu(B) = \frac{1}{2\pi} \int_0^{2\pi} \chi_B(e^{it}) \, dt \, , \tag{7.4}$$

holds, where χ_B denotes the characteristic function of the set B.

Proof. Our statement follows from the simple observation that the mapping $t \mapsto e^{it}$ is bijective and continuous from $[0, 2\pi[$ onto $\mathbb{T}$, which maps Borel sets onto Borel sets. Hence the set function, defined by the right hand side of the formula (7.4) on the Borel sets in $\mathbb{T}$, and obviously possesses the properties of the Haar measure on $\mathbb{T}$, is actually identical to the Haar measure, by uniqueness. □

We note that, in addition to the statement in the previous theorem, the interval $[0, 2\pi]$ equipped with addition modulo 2π and with the Euclidean topology, then, after identifying 0 and 2π, we obtain a compact Abelian group topologically isomorphic to $\mathbb{T}$, where the isomorphism is provided by the mapping $t \mapsto e^{it}$. As a result, in what follows we shall identify these two groups. Moreover, this means that the continuous functions defined on $\mathbb{T}$ can be identified with those continuous functions on the interval $[0, 2\pi]$, which take the same value at the endpoints of this interval. We can move one step further, by identifying these functions with those defined, continuous, and periodic by 2π on the whole real line. In the background of this identification we find the simple fact that the factor group $\mathbb{R}/\mathbb{Z}$ of the topological group $\mathbb{R}$ with respect to its closed subgroup $\mathbb{Z}$ equipped with the factor topology is a topological group isomorphic to $\mathbb{T}$.

7.3 Fourier series

Harmonic analysis on the group $\mathbb{T}$ is identical to the classical theory of Fourier series for periodic functions. Here we give a short summary of the results, which are consequences of the above considerations in this respect. We apply the ideas used in Section 2.1.

We have seen in Theorem 6.2 that the dual of $\mathbb{T}$ is topologically isomorphic to $\mathbb{Z}$, and the isomorphism is obtained in the following way: the elements of $\mathbb{T}$ can be identified by the complex numbers of the form e^{it}, where t is in $[0, 2\pi]$. Then each character of $\mathbb{T}$ has the form

$$\chi_n(e^{it}) = e^{int}$$

where n is an integer, and the correspondence $\chi_n \longleftrightarrow n$ is the natural isomorphism between $\widehat{\mathbb{T}}$ and $\mathbb{Z}$. We identify the complex valued functions on $\mathbb{T}$ with the 2π periodic functions on $\mathbb{R}$. The space $L^2(\mathbb{T})$ can be identified with the space $L^2[0, 2\pi]$, hence if $f, g : \mathbb{T} \to \mathbb{C}$ are given functions in $L^2(\mathbb{T})$, then their inner product is

$$\langle f, g\rangle = \frac{1}{2\pi}\int_0^{2\pi} f(e^{it})\,\overline{g(e^{it})}\,dt\,.$$

The next theorem follows immediately.

Theorem 7.6. *The function sequence $(e^{int})_{n\in\mathbb{Z}}$ is a complete orthonormal system in $L^2[0, 2\pi]$.*

Proof. Let $\chi_n(e^{it}) = e^{int}$, whenever n is an integer and t is a real number. Then we have for $n \neq m$

$$\langle \chi_n, \chi_m\rangle = \frac{1}{2\pi}\int_0^{2\pi} e^{int}\,\overline{e^{imt}}\,dt = \frac{1}{2\pi}\int_0^{2\pi} e^{i(n-m)t}\,dt$$

$$= \frac{1}{2\pi i(n-m)}\left[e^{i(n-m)t}\right]_0^{2\pi} = 0\,,$$

while this inner product is 1 for $n = m$. This proves the orthonormality of the function system consisting of all characters. The completeness is a consequence of the Stone–Weierstrass Theorem. Indeed, $\mathbb{T}$ is a compact Hausdorff space, and the linear combinations of the characters, the so-called *trigonometric polynomials*, form a separating subalgebra in the algebra of all continuous complex functions, which contains the identically 1 function and is closed under complex conjugation. Hence, by the complex version of the Stone–Weierstrass Theorem 3.57, these functions form a dense set with respect to uniform convergence in the space of all continuous functions. As uniform convergence obviously implies L^2-convergence, and the continuous functions form a dense set in $L^2(\mathbb{T})$, hence the trigonometric polynomials form a dense set in $L^2(\mathbb{T})$, too. It follows that if a function in $L^2(\mathbb{T})$ is orthogonal to each character, then it is orthogonal to their linear combinations, hence to all trigonometric polynomials, and by their density, to each element in $L^2(\mathbb{T})$. We infer that it is orthogonal to itself, hence it must be zero. □

The function system $(e^{int})_{n\in\mathbb{Z}}$ is called the *complex trigonometric system*, or simply the *trigonometric system.* Similarly as we did it in Section 2.1, we associate to each function in $L^2(\mathbb{T})$ the sequence of Fourier coefficients in its expansion with respect to the trigonometric system: for each function f in $L^2(\mathbb{T})$ and integer n we let

$$c_n = \widehat{f}(n) = \frac{1}{2\pi}\int_0^{2\pi} f(e^{it})\, e^{-int}\, dt\,. \tag{7.5}$$

The complex valued function $\widehat{f}$ defined on $\mathbb{Z}$ is the *Fourier transform* of f, which can be identified with the complex sequence $(c_n)_{n\in\mathbb{Z}}$, whose terms are called the *Fourier coefficients* of f. Using this notation, by the completeness of the trigonometric system, we have for each function f in $L^2(\mathbb{T})$

$$f = \sum_n \widehat{f}(n)\, \chi_n\,, \tag{7.6}$$

which is the analogue of the Inversion Theorem 2.2. Obviously, the convergence of the series is meant in the Hilbert space $L^2(\mathbb{T})$. From Hilbert space theory it is known that the identity

$$\frac{1}{2\pi}\int_0^{2\pi} |f(e^{it})|^2\, dt = \sum_n |\widehat{f}(n)|^2 = \sum_n |c_n|^2 \tag{7.7}$$

holds, which is the analogue of Parseval's Formula (2.1).

Equation (7.6) can also be written in the form

$$f(e^{it}) = \int \widehat{f}(n)\, e^{int}\, dn\,, \tag{7.8}$$

where dn is the counting measure on $\mathbb{Z}$. If the space of square integrable functions corresponding to the counting measure on $\mathbb{Z}$ is denoted by $L^2(\mathbb{Z})$, then (7.7) expresses the property that for each f in $L^2(\mathbb{T})$ the function $\widehat{f}$ belongs to the space $L^2(\mathbb{Z})$, and (7.7) can be written in the form

$$\|f\| = \|\widehat{f}\|\,. \tag{7.9}$$

Here we summarize our results in the following theorem, which corresponds to the theorem of Plancherel 2.1.

Theorem 7.7. *(Plancherel's Theorem) On the compact Abelian group $\mathbb{T}$ the Fourier transformation is an isometric isomorphism of $L^2(\mathbb{T})$ onto $L^2(\mathbb{Z})$.*

Proof. The surjectivity of the Fourier transformation, which is the only statement here we have not yet proved, is a consequence of the general theory of Hilbert spaces. □

As it is easy to see, the measure dn has almost all basic properties of the Haar measure on compact Abelian groups: the only exception is the finiteness, as its value is finite exactly on the finite sets, or, in other words, on those sets, which are compact in the discrete topology of $\mathbb{Z}$. We shall see that we have a similar situation concerning some other compact, and even non-compact, locally compact Abelian groups and their duals: both the group G and its dual $\widehat{G}$ can be equipped with Borel measures, which are translation invariant such that there is an isometric isomorphism between the corresponding $L^2(G)$ and $L^2(\widehat{G})$ Hilbert spaces, having similar properties to those of the Fourier transformation. This fundamental property of the Fourier transformation makes it a basic tool of harmonic analysis. Our main goal in the forthcoming investigations is to formulate and prove this theorem in more and more general circumstances.

7.4 Fourier analysis on compact elementary Abelian groups

The ideas in the previous section can be extended to compact Abelian groups, which are definitely much more general than the complex unit circle. Which is the most general compact Abelian group? If G is a compact Abelian group, then, by the Duality Theorem, we have $G = \widehat{\widehat{G}}$. Here $\widehat{G}$ is a discrete Abelian group, hence each compact Abelian group is the dual of some discrete Abelian group. In other words, every compact Abelian group can be identified with a group consisting of some functions defined on a certain discrete Abelian group with values in the complex unit circle, equipped with a reasonable topology and with the pointwise multiplication. This function space is nothing but a closed subspace of a direct power of the compact topological space $\mathbb{T}$, and the group operation is the pointwise multiplication of functions. Roughly speaking, we can imagine every compact Abelian group as a set of certain functions on a discrete Abelian group, which take complex numbers of modulus one, and the group operation is the pointwise multiplication. However, as every discrete Abelian group is the dual of some compact Abelian group, hence to achieve a general overlook on all compact Abelian groups we should know all discrete Abelian groups, a dream of algebraists, which is far from reality at this moment. Nevertheless, we may apply the reverse argument: all discrete Abelian groups, which are basically defined by a set of abstract axioms, are nothing but sets of continuous functions defined on some compact topological spaces, taking complex numbers of modulus 1 as their values, and this structure is equipped with pointwise multiplication as group operation. It is only a question of taste which is the primary concept: discrete Abelian groups, defined purely axiomatically, or continuous functions defined on compact topological spaces, which are, by the way, defined by axioms, too. We do not commit ourselves in this question, we just call the reader's attention that the more discrete Abelian groups can be described by purely algebraic tools the more compact Abelian groups are becoming well-known for us, as their duals. And conversely, discovering analytic and topological properties of wider and wider classes of compact Abelian groups results in, as a reward, a deeper insight into the structure of more and more discrete Abelian groups.

Now we shall have a look at the elementary compact Abelian groups. In Section 6.2 we have seen that these groups have the form $G = \mathbb{T}^m \times F$, where m is a nonnegative integer, and F is a finite Abelian group. We know that F is a direct sum of finitely many finite cyclic groups, which have the form $\mathbb{Z}/n\mathbb{Z}$. In what follows the group $\mathbb{T}$ will be identified with $[0, 2\pi]$, equipped with the euclidean topology and with addition modulo 2π, hence it becomes a compact Abelian group. Moreover, F is considered as a direct sum of finitely many copies of the integers equipped with the discrete topology and with addition modulo finitely many arbitrary nonzero integers, that is

$$F = \mathbb{Z}/n_1\mathbb{Z} \oplus \mathbb{Z}/n_2\mathbb{Z} \oplus \cdots \oplus \mathbb{Z}/n_l\mathbb{Z}\,,$$

where $n_1, n_2, \ldots, n_l$ are nonnegative integers. Hence, in this sense, the group G is a direct sum of m copies of the "continuous circle", and l copies of "finite circles". Functions on G can be realized as functions of $m + l$ variables, the first m are reals variables, and the function is 2π periodic in these variables, while the remaining l variables take integer values, further the function is periodic with respect to $n_1, n_2, \ldots, n_l$, respectively, in these integer variables. Denoting the variable by (x, k) we can write in coordinate-form

$$(x, k) = (x_1, x_2, \ldots, x_m, k_1, k_2, \ldots, k_l)\,,$$

where $x_1, x_2, \ldots, x_n$ are real numbers, and $k_1, k_2, \ldots, k_l$ are integers. The Haar measure on G is the product of the Haar measures on the factors, that is, $d\lambda^m \times ds_1 \times ds_2 \times \cdots \times ds_l$, where $d\lambda$ is the Lebesgue[5] measure on the interval $[0, 2\pi]$ multiplied by $\frac{1}{2\pi}$, and ds_j is the counting measure on $\mathbb{Z}/\, n_j\mathbb{Z}$, as a set of n_j elements, multiplied by $\frac{1}{n_j}$. We let $n = n_1 \cdot n_2 \cdot \cdots \cdot n_l$, the number of elements of the group F.

The inner product of the functions $f, g : G \to \mathbb{C}$ in $L^2(G)$ is

$$\langle f, g\rangle = \frac{1}{2^m \pi^m n} \sum_{k_j=0}^{n_j-1} \sum_{k_2=0}^{n_2-1} \cdots \sum_{k_l=0}^{n_l-1} \int_{[0,2\pi]^m} f(x,k)\overline{g(x,k)}\, d\lambda^m(x)\,. \quad (7.10)$$

In particular

$$\|f\|^2 = \frac{1}{2^m \pi^m n} \sum_{k_j=0}^{n_j-1} \sum_{k_2=0}^{n_2-1} \cdots \sum_{k_l=0}^{n_l-1} \int_{[0,2\pi]^m} |f(x,k)|^2\, d\lambda^m(x)\,.$$

One can see from these formulas that in the first variable we compute the integral mean, and in the second one the arithmetical mean, separately in each component. We shall abbreviate integration with respect to the Haar measure on F, that is the inner summation in the above equation, simply by $\sum_{k \in F}$. Hence, by Theorems 1.6 and 6.4, the dual of G is

H. L. Lebesgue

$$\widehat{G} = \mathbb{Z}^m \times \mathbb{Z}/n_1\mathbb{Z} \times \mathbb{Z}/n_2\mathbb{Z} \times \cdots \times \mathbb{Z}/n_l\mathbb{Z}\,,$$

and the measure on $\widehat{G}$, corresponding to the Haar measure on G, is, by the above consideration,

$$dn^m \times ds_1 \times ds_2 \times \cdots \times ds_l\,, \quad (7.11)$$

because, as we have seen it in section 2.1, this choice results in the well-known form of the Inversion Theorem. The Fourier transform of the function $f : G \to \mathbb{C}$ in $L^2(G)$ is $\hat{f} : \mathbb{Z}^m \times F \to \mathbb{C}$ is defined by

$$\hat{f}(p,q) = \frac{1}{2^m \pi^m n} \sum_{k \in F} \int_{[0,2\pi]^m} f(x,k) e^{-(x \bullet p + k \bullet q)}\, d\lambda^m(x)\,, \quad (7.12)$$

[5]Henry Léon Lebesgue, French mathematician (1875-1941)

where p is in $\mathbb{Z}^m$ and q is in F, further $u \bullet v$ denotes the inner product in $\mathbb{R}^m$ and in $\mathbb{R}^n$. Plancherel's Theorem takes the following form:

Theorem 7.8. *Let G be an elementary compact Abelian group. The Fourier transformation, given by formula (7.12), is an isometric isomorphism of $L^2(G)$ onto $L^2(\widehat{G})$, and its inverse is given by*

$$f(x,k) = \sum_{p\in\mathbb{Z}^m} \sum_{q\in F} \hat{f}(p,q)\, e^{i(x\bullet p+k\bullet q)}\,, \tag{7.13}$$

whenever $G = \mathbb{T}^m \times F$ has the above form, f is in $L^2(G)$, x is in $\mathbb{T}^m$, and k is in F.

Proof. As the Hilbert space $L^2(G)$ is the direct sum of m copies of $L^2(\mathbb{T})$ and of the Hilbert space $L^2(F)$, hence it is enough to prove our statement in the case $G = F$ and $G = \mathbb{T}$. The case $G = F$ was treated in Section 2.1, and the case $G = \mathbb{T}$ in Section 7.3, hence our statement is proved. □

The equality in the above formula is obviously meant in the sense of $L^2(G)$ convergence. We note that this theorem corresponds to the theorem of F. Riesz and E. S. Fischer[6], which is well-known in Hilbert space theory (see [Dunford and Schwartz (1988a)], Chapter IV. pp. 251–256). The mapping $p \mapsto \hat{f}(p,q)$ is a "sequence" labeled by the elements of $\mathbb{Z}^m$ for each q in F, which can be considered as a generalization of the sequence of classical Fourier coefficients. As the elements of F are vectors of n nonnegative integer components, this "sequence-vector" is in l^2 in the sense that we have

$$\sum_{p\in\mathbb{Z}^m} \sum_{q\in F} |\hat{f}(p,q)|^2 < \infty\,,$$

which corresponds to Parseval's Formula (2.1). The space $L^2(\widehat{G})$ is formed by all these "sequence-vectors". We define for each p in $\mathbb{Z}^m$, k, q in F and real x

$$\chi_{p,q}(x,k) = e^{i(x\bullet p+k\bullet q)}\,,$$

then the system of functions $(\chi_{p,q})_{p\in\mathbb{Z}^m, q\in F}$ corresponds to the trigonometric system, and, by our former results, it is a complete orthonormal system in $L^2(G)$, as it follows easily from Theorem 5.8, and from the simple fact that these functions are exactly the characters of the direct sum $\mathbb{T}^m \oplus F$. Hence the Inversion Formula (7.13) is exactly the representation of f in this orthonormal basis.

[6]Ernst Sigismund Fischer, Austrian mathematician (1875-1954)

7.5 Fourier analysis on compact Abelian groups

The results in the previous section can be extended to arbitrary compact Abelian groups. The key is Theorem 5.8. The only difficulty is that the structure of general compact Abelian groups cannot be described by using direct sums or products of "simpler" Abelian groups. Although the dual of a compact Abelian group is always discrete, hence the orthonormal basis of $L^2(G)$ consisting of all characters is also a "discrete" set, however, if $L^2(G)$ is not separable, then this discrete set is not countable, hence it cannot be ordered in a sequence, consequently, the Fourier transform is neither a sequence, too. Consequently, it is not clear what is the corresponding Inversion Formula in this case. Nevertheless, it is possible to make the sum on the right of equation (7.13) meaningful.

Let X be a set, H a Hilbert space, h an element of H, and $\varphi : X \to H$ a function. We say that $\sum_{x \in X} \varphi(x)$ *is convergent, and its sum is* h, in symbols

$$\sum_{x \in X} \varphi(x) = h\,,$$

if for each $\varepsilon > 0$ there is a finite subset $X_0 \subseteq X$ such that if $X' \supseteq X_0$ is finite, then

$$\|h - \sum_{x \in X'} \varphi(x)\| < \varepsilon\,.$$

In this case we simply say that the series $\sum_{x \in X} \varphi(x)$ is convergent, and its sum is h. It is easy to see that in the case $X = \mathbb{N}$ this is equivalent to the absolute convergence of series. It is also easy to check that if the series $\sum_{x \in X} \varphi(x)$ is convergent, then for each $\varepsilon > 0$ there exists a finite subset X_0 in X such that whenever the finite subsets X_1, X_2 of X contain X_0, then

$$\|\sum_{x \in X_1} \varphi(x) - \sum_{x \in X_2} \varphi(x)\| < \varepsilon\,.$$

It follows easily that for each positive integer n the set

$$\{x \in X : \|f(x)\| \geqslant \frac{1}{n}\} \tag{7.14}$$

is finite, hence for every convergent series $\sum_{x \in X} \varphi(x)$ the *support* of the function φ, that is, the set of all x in X with $\varphi(x) \neq 0$, is countable.

The following theorem can be proved by using well-known techniques from Hilbert space theory.

Theorem 7.9. *Let H be a Hilbert space, and let $\mathcal{E}$ be a complete orthonormal set in H. Then we have the following statements.*

1. *For each x in H the series $\sum_{e\in\mathcal{E}}\langle x,e\rangle e$ is convergent, and its sum is x:*
$$x=\sum_{e\in\mathcal{E}}\langle x,e\rangle e\,.$$
2. *For each x,y in H the series $\sum_{e\in\mathcal{E}}\langle x,e\rangle\overline{\langle y,e\rangle}$ is convergent, and its sum is $\langle x,y\rangle$:*
$$\langle x,y\rangle=\sum_{e\in\mathcal{E}}\langle x,e\rangle\overline{\langle y,e\rangle}\,.$$
3. *For each x in H the series $\sum_{e\in\mathcal{E}}|\langle x,e\rangle|^2$ is convergent, and its sum is $\|x\|^2$:*
$$\|x\|^2=\sum_{e\in\mathcal{E}}|\langle x,e\rangle|^2\,.$$

Let G be a compact Abelian group, and let $c:\widehat{G}\to\mathbb{C}$ be a function. Applying the above concepts to the space $L^2(G)$ we say that the series $\sum_{\chi\in\widehat{G}}c(\chi)\chi$ *is convergent in* $L^2(G)$, and its sum is the function $f:G\to\mathbb{C}$ in $L^2(G)$, if this holds in the previous sense, that is, for each $\varepsilon>0$ there is a finite subset H_0 in $\widehat{G}$ such that for each finite set $H\supseteq H_0$ we have
$$\|f-\sum_{\chi\in H}c(\chi)\chi\|<\varepsilon\,.$$
In this case we use the notation
$$f=\sum_{\chi\in\widehat{G}}c(\chi)\chi\,.$$
In particular, if $f:G\to\mathbb{C}$ is a function in $L^2(G)$, then its *Fourier transform* is defined by
$$\hat{f}(\chi)=\int f(x)\overline{\chi}(x) \tag{7.15}$$
for each character χ. Hence $\hat{f}:\widehat{G}\to\mathbb{C}$ is a function defined on the dual of G. Actually, the values of the function $\hat{f}$ are nothing but the analogues of the classical Fourier coefficients: $\hat{f}(\chi)$ is exactly the component of the element f with respect to the basis vector χ in the orthonormal basis of the Hilbert space $L^2(G)$ consisting of characters. The following theorems are immediate consequences of this simple observation.

Theorem 7.10. *(Parseval's Formula) Let G be a compact Abelian group. For each f in $L^2(G)$ we have*
$$\|f\|^2=\sum_{\chi\in\widehat{G}}|\hat{f}(\chi)|^2\,. \tag{7.16}$$

Theorem 7.11. *(Plancherel's Theorem) Let G be a compact Abelian group. The Fourier transformation is an isometric isomorphism of $L^2(G)$ onto $L^2(\widehat{G})$, and its inverse is given by*

$$f = \sum_{\chi \in \widehat{G}} \hat{f}(\chi)\chi \,. \tag{7.17}$$

This theorem obviously generalizes those we have proved in the cases of finite and of elementary compact Abelian groups. One can see that the measure on the dual of a compact Abelian group corresponding to the Haar measure, which ensures the validity of the Inversion Formula in the form (7.17) is exactly the counting measure. If we choose another constant multiple of the Haar measure on G, then for the validity of (7.17) we have to multiply the counting measure on $\widehat{G}$ with some suitable factor, too.

7.6 Integrable functions on compact Abelian groups

The existence of the Haar measure and the Haar integral makes it possible to define the Fourier transform not just for functions in L^2, but for all integrable functions. Moreover, this extension is possible for the major part of Lebesgue spaces from the theory of measure and integral, even on non-compact Abelian groups. If m denotes the Haar measure on the compact Abelian group G, then we define the corresponding Lebesgue space $L^p(G)$ for $p \geqslant 1$ in the usual way: identifying the functions which are equal almost everywhere with respect to m the measurable function $f : G \to \mathbb{C}$ belongs to $L^p(G)$, if

$$\|f\|_p = \left[\int |f|\, dm\right]^{1/p} < +\infty \,.$$

It is well-known that, equipped with this norm, further with the pointwise addition and multiplication by scalars, $L^p(G)$ is a Banach space. We extend this to the case $p = +\infty$, too, when defining the *essential supremum* $\|f\|_\infty$ of the measurable function f by

$$\text{ess sup}\,|f| = \inf\{a \in \mathbb{R} : \mu(\{x : |f(x)| > a\}) = 0\} \,. \tag{7.18}$$

The space $L^\infty(G)$ of all *essentially bounded functions* is the set of those measurable functions for which this quantity is finite. In the sequel we shall make use of the following basic inequality (see e.g. [Hewitt and Ross (1979)], (12.4) Theorem, p. 137).

Theorem 7.12. *(Hölder's Inequality) Let $(X, \mathcal{A}, \mu)$ be a measure space, p, q positive real numbers with $p^{-1} + q^{-1} = 1$, and let $f, g : X \to \mathbb{C}$ be functions with f in $L^p(X)$ and g in $L^q(X)$. Then $f \cdot g$ is integrable, and we have*

$$\int_X |f \cdot g| \, d\mu \leqslant \|f\|_p \cdot \|g\|_q \, .$$

The same conclusion holds for $p = 1$ with $q = +\infty$, and for $p = +\infty$ with $q = 1$.

Since in the case of compact G the function 1 is integrable, by Hölder's[7] Inequality 7.12, we have $L^p(G) \subseteq L^1(G)$, whenever $1 \leqslant p \leqslant +\infty$. It is also clear that for each character χ and f in $L^p(G)$ the function $f \cdot \chi$ belongs to $L^p(G)$, too, hence the Fourier transform of f can be defined by the usual formula

$$\hat{f}(\chi) = \int f \cdot \overline{\chi} \, dm$$

for each character χ.

Besides the linear ones there is another important operation defined on $L^1(G)$, namely the respective one introduced in Section 2.2 on finite groups: the convolution. We shall consider it in the more general case of locally compact Abelian groups.

7.7 Translation invariant spaces

Suppose that G is a compact Abelian group. Then, by the results of Section 7.5, every function in $L^2(G)$ is determined by its Fourier coefficients – at least in L^2-sense: functions, which differ only on a set of zero Haar measure are indistinguishable. In particular, every continuous function on G is completely determined by its Fourier coefficients, because the properties of the Haar measure obviously imply that the complement of a set with zero Haar measure is dense, and continuous functions, which coincide on dense sets are identical. Hence continuous functions are completely determined by those characters, which actually take part in its harmonic analysis in the sense, that the corresponding Fourier coefficients of the function are different from zero. It turns out that these characters are exactly the ones in the translation invariant closed subspace generated by the function. Now we have a closer look at this problem.

[7] Otto Ludwig Hölder, German mathematician (1859-1937)

Let G be a compact Abelian group. Then the space $\mathcal{C}(G)$ is a Banach space, when equipped with the norm $\|f\| = \sup_{x\in G}|f(x)|$. Convergence in this Banach space is the uniform convergence. A subspace is called a *variety*, if it is closed and translation invariant. In other words, a linear subspace of $\mathcal{C}(G)$ is a variety if and only if it is closed under uniform convergence, and it contains all translates of its elements. For a subset H in $\mathcal{C}(G)$ the intersection of all varieties containing H is obviously a variety, which is called the variety *generated by* H, and it is denoted by $\tau(H)$. In particular, if H is a singleton, $H = \{f\}$, then we write $\tau(f)$ for the generated variety, and we call it the *variety of* f. We show that varieties are actually closed under all convolution operators. Here we use a slightly different meaning of these operators from that we had in Section 5.4. From Theorem 7.3 it follows that the dual space of $\mathcal{C}(G)$ can be identified with $\mathcal{M}(G)$, the space of all regular Borel measures on G. For each μ in $\mathcal{M}(G)$ and f in $\mathcal{C}(G)$ the mapping $f \mapsto \mu * f$ is a continuous linear operator on $\mathcal{C}(G)$, as it is easy to see. Such operators are called *convolution operators.*

Theorem 7.13. *Let G be a compact Abelian group and V a variety in $\mathcal{C}(G)$. Then V is invariant under every convolution operator. Conversely, every closed linear subspace of $\mathcal{C}(G)$, which is invariant under all convolution operators, is a variety.*

Proof. Let f be in V and μ be in $\mathcal{M}(G)$. To show that $\mu * f$ is in V it is enough to show, by Theorem 3.50, that every linear functional of $\mathcal{C}(G)$, which vanishes on V, also vanishes on $\mu * f$. In other words, we suppose that ν is in $\mathcal{M}(G)$ and ν annihilates V. Then $\tau_y f$ is in V hence

$$\nu(\mu * f) = \int (\mu * f)(x)\, d\nu(x) = \int\int f(x-y)\, d\mu(y)\, d\nu(x)$$

$$= \int\Big(\int \tau_y f(x)\, d\nu(x)\Big)\, d\mu(y) = 0\,.$$

Conversely, if V is a closed linear subspace, which is closed under all convolution operators, then for each f in V and y in G we have that $\tau_y f = \delta_{-y} * f$ is in V. □

Obviously, each operator of the form $f \mapsto a * f$ with a is continuous is a convolution operator.

Theorem 7.14. *Let G be a compact Abelian group, and let f be a continuous function on G. Then the variety of f contains exactly those characters, which are not zeros of the Fourier transform of f.*

Proof. First we show that a character χ is in $\tau(f)$ if and only if $\widehat{f}(\chi) \neq 0$. Indeed, we have for each character χ

$$(f * \chi)(x) = \int \chi(x-y) f(y)\, dy = \widehat{f}(\chi)\chi(x)\,.$$

It follows that each character χ for which $\widehat{f}(\chi) \neq 0$, is a constant multiple of $\chi * f$, hence belongs to $\tau(f)$, by the previous theorem. Conversely, if $\widehat{f}(\chi) = 0$, then we have $\langle f, \chi \rangle = 0$ and for each y it follows

$$\langle \tau_{-y} f, \chi \rangle = \int f(x+y)\overline{\chi(y)}\, dy = \chi(x) \int \chi(-y) f(y)\, dy = 0\,,$$

hence χ is orthogonal to $\tau(f)$, which implies that it is not in $\tau(f)$. □

This theorem enlightens the role of the Fourier transform of f: it selects those characters, which actually take part in the harmonic analysis of f by assigning them nonzero coefficients. Those characters, which are zeros of the Fourier transform do not play any role in the reconstruction process of f from the basic functions of its variety. However, the reconstruction process, the *synthesis* of f is guaranteed at this moment in the L^2-sense only: Plancherel's Theorem 7.11 asserts the L^2-convergence of the Fourier series. In the theory of classical Fourier series it is a distinguished area to discuss the pointwise, and also the uniform convergence of Fourier series. Without going into the details we just mention here, that the famous theorems of L. Fejér[8] establish the uniform convergence of the arithmetical means of the partial sums of the Fourier series for each continuous function, which means that the linear span of all characters in $\tau(f)$ is always dense in $\tau(f)$. The interested reader will find references on pointwise and uniform convergence of Fourier series e.g. in [Edwards (1982)], [Edwards (1979)], [Hewitt and Ross (1970)]. Relations to the theory of almost periodic functions are also worth mentioning (see e.g. [Maak (1967)]). A detailed investigation of spectral analysis and spectral synthesis for varieties is the subject of the second part of this book.

[8]Lipót Fejér, Hungarian mathematician (1880-1959)

Chapter 8

DUALITY OF LOCALLY COMPACT ABELIAN GROUPS

8.1 Compactly generated locally compact Abelian groups

The locally compact Abelian group G is compactly generated, if the unit element has a neighborhood with compact closure, which neighborhood *generates* G. If G is a compactly generated locally compact Abelian group, then it is easy to see that the identity element has a symmetric neighborhood with compact closure, which generates G.

Theorem 8.1. *In a compactly generated locally compact Abelian group every subgroup generated by a single element, is either discrete or it has a compact closure.*

Proof. Let g be an element of G, and let H denote the closed subgroup generated by g. Suppose that H is not discrete. If U is a symmetric neighborhood in H, then for each natural number N there exists a natural number $n > N$ such that g^n is in U. Indeed, supposing the contrary U contains only a finite number of powers of g, hence there is a neighborhood of the identity e, which does not contain any power of g with nonzero exponent, hence H is discrete.

If V is a nonempty open subset of H, then V contains some power of g with positive exponent. Indeed, there is an integer l for which g^l is in V, as the powers of g form a dense subset in H. Then there is a symmetric neighborhood U of the identity in H such that $g^l U \subseteq V$, and there is a natural number $N > |l|$ such that g^N is in U, hence g^{l+N} belongs to V and $l + N > 0$.

Let W be a symmetric neighborhood of the identity with compact closure in H. Then for arbitrary x in H the element g^k belongs to xW for some positive integer k, hence x is in g^kW. It follows that the sets g^kW, where k is a natural number, form a covering of H, hence of W^{cl}, too. The latter set is compact, hence there is a positive integer N such that $W^{cl} \subseteq gW \cup \cdots \cup g^N W$ holds. If x is in H, then let $m(x)$ denote the smallest positive integer for which $g^{m(x)}$ belongs to xW. Then $x^{-1}g^{m(x)}$ belongs to W, hence there exists an integer $0 < j \leqslant N$ such that $x^{-1}g^{m(x)}$ is in g^jW, that is, $g^{m(x)-j}$ is in xW. This means $m(x) - j \leqslant 0$, hence $0 < m(x) \leqslant j \leqslant N$, and x belongs to the set $g^{m(x)}W$. In other words, $H \subseteq gW \cup \cdots \cup g^N W$ holds, and W^{cl} is compact, hence H is a compact set, too. □

Theorem 8.2. *Compactly generated locally compact Abelian group has a finitely generated discrete subgroup, for which the corresponding factor group is compact.*

Proof. Let U be a symmetric neighborhood of the identity with compact closure, which generates G. As $(U^{cl})^2$ is compact, hence there are elements $a_1, a_2, \ldots, a_n$ in G such that $(U^{cl})^2 \subseteq a_1U \cup a_2U \cup \cdots \cup a_nU$ holds. Let A be the subgroup generated by the elements $a_1, a_2, \ldots, a_n$. As $U^2 \subseteq AU$, hence we have

$$U^3 \subseteq AU^2 \subseteq A^2U = AU\,,$$

and similarly, $U^n \subseteq AU$, therefore $G = \cup_{n=1}^{\infty} U^n \subseteq AU$, that is $G = AU$. If the closure of each subgroup generated by every a_i is compact, then G is compact, and the proof is complete. Hence, by the previous theorem, there is an a_i, which generates a discrete subgroup in G; suppose that b_1 is one of them. Assume that we have selected the elements $b_1, b_2, \ldots, b_i$ out of $a_1, a_2, \ldots, a_n$ with the property that they generate a discrete subgroup N_i in G. We show that if the factor group G/N_i is not compact, then we can continue with selecting an element b_{i+1} out of $a_1, a_2, \ldots, a_n$ with the above property. If $\Phi : G \to G/N_i$ is the natural homomorphism, then $G/N_i = \Phi(G) = \Phi(A)\Phi(U)$, and if G/N_i is not compact, then at least one of the elements $\Phi(a_1), \Phi(a_2), \ldots, \Phi(a_n)$ generates a discrete subgroup in G/N_i; let $\Phi(a_j)$ be one of them, further let $b_{i+1} = a_j$. Then it is easy to see that $b_1, b_2, \ldots, b_i, b_{i+1}$ generate a discrete subgroup in G. As A is finite, this process terminates after a finite number of steps. The proof is complete. □

If G is a locally compact Abelian group, and H is a closed subgroup in G, then the *annihilator* of H in $\widehat{G}$ is the set of those characters $A(\widehat{G}, H)$, which take the value 1 at the elements of H. The introduction of annihilators makes it possible to turn algebraic properties into topological ones and vice versa. We shall utilize this observation in the sequel, in particular, when proving the basic approximation theorem in the following section.

Theorem 8.3. *The annihilator of every closed subgroup in a locally compact Abelian group is a closed subgroup.*

Proof. It is clear, by the definition of the compact-open topology. □

Theorem 8.4. *Let G be a discrete or compact Abelian group, and let H be a closed subgroup in G. Then we have*

1. $(G/H)\hat{}= A(\widehat{G}, H)$;
2. $A(G, A(\widehat{G}, H)) = H$.

Proof. The proof of the first statement is the following: let Φ denote the natural homomorphism of G onto G/H. We show that the mapping $\psi \mapsto \psi \circ \Phi$ is a topological isomorphism of $(G/H)\hat{}$ onto $A(\widehat{G}, H)$. It is clear that for each character ψ of G/H the function $\psi \circ \Phi$ is a character of G, which takes the value 1 at the elements of H. Obviously, $\psi \mapsto \psi \circ \Phi$ is a homomorphism. If $\psi \circ \Phi$ is the identity of $A(\widehat{G}, H)$, then $\psi(\Phi(x)) = 1$ holds for each x in G, hence ψ is the identity of $(G/H)\hat{}$, thus the mapping $\psi \mapsto \psi \circ \Phi$ is an isomorphism of $(G/H)\hat{}$ into $A(\widehat{G}, H)$. If φ is a character of G, identically 1 on H, then $\Phi(x) = \Phi(y)$ implies $\Phi(xy^{-1}) = e$, hence xy^{-1} belongs to H, which means $\varphi(xy^{-1}) = 1$, hence $\varphi(x) = \varphi(y)$. It follows that if we define ψ by $\psi(xH) = \varphi(x)$, then ψ is a character of G/H and we have $\psi \circ \Phi = \varphi$. Consequently, the mapping $\psi \mapsto \psi \circ \Phi$ is surjective, further, it is obviously continuous. If G is compact, then $\widehat{G}$ is discrete, and so is $A(\widehat{G}, H)$, which implies the continuity of the inverse mapping. On the other hand, if G is discrete, then $\widehat{G}$ is compact, and so is $A(\widehat{G}, H)$, further the inverse of a continuous bijective mapping on a compact Hausdorff space is continuous, by Theorem 3.15.

Now we prove the second statement. It is clear that $H \subseteq A(G, A(\widehat{G}, H))$, as for each h in H and x in $A(\widehat{G}, H)$ it follows $h(x) = 1$, by the definition of the annihilator. Suppose that x is an element of $A(G, A(\widehat{G}, H))$, which is not in H. Then there is a character χ of G such that $\chi(x) \neq 1$ holds, and χ is identically 1 on H. Hence χ belongs to $A(\widehat{G}, H)$, and for each x in H we have $\chi(x) = 1$, which is a contradiction. □

8.2 The approximation theorem of compactly generated locally compact Abelian groups

In the following two theorems we prepare the famous Approximation Theorem of Pontryagin.

Theorem 8.5. *In every compact Abelian group each neighborhood of the identity contains a closed subgroup such that the corresponding factor group is an elementary group.*

Proof. Let U be a neighborhood of the identity in G. As $G = \widehat{\widehat{G}}$, there is a real number $\varepsilon > 0$ and there are elements $\chi_1, \chi_2, \ldots, \chi_n$ in $\widehat{G}$ such that the set of those elements x in G for which $|\chi_i(x) - 1| < \varepsilon$ holds whenever $i = 1, 2, \ldots, n$, is a subset of U. Let H be the subgroup in $\widehat{G}$ generated by the elements $\chi_1, \chi_2, \ldots, \chi_n$. Then the closed subgroup N of those elements x in G at which all elements of H take the value 1, that is $A(G, H)$, is a subset of U. Moreover $H = A(\widehat{G}, N) = (G/N)\widehat{}$ is finitely generated, hence $(G/N)\widehat{} = \mathbb{Z}^k \times F$ holds with some nonnegative integer k and finite Abelian group F. It follows that $G/N = (G/N)\widehat{\widehat{}} = (\mathbb{Z}^k \times F)\widehat{} = \mathbb{T}^k \times F$ is an elementary group. □

Theorem 8.6. *If the locally compact Abelian group G has a finitely generated discrete subgroup such that the corresponding factor group is a compact elementary group, then G is an elementary group.*

Proof. Let N be a finitely generated discrete subgroup of G such that the corresponding factor group is a compact elementary group: $G/N = \mathbb{T}^r \times F$, where r is a nonnegative integer, and F is a finite Abelian group.

First we assume that G is connected. Then G/N is connected, too, hence $G/N = \mathbb{T}^r = \mathbb{R}^r/\mathbb{Z}^r$. Let ψ, resp. φ be the natural homomorphism of $\mathbb{R}^r$ onto $\mathbb{T}^r$, resp. of G onto $\mathbb{T}^r$. As $\mathbb{Z}^r$, resp. N is a discrete subgroup of $\mathbb{R}^r$, resp. of G, hence there is a neighborhood U of 0 in $\mathbb{R}^r$ and a neighborhood V of the unit in G such that ψ is one-to-one on U, and φ is one-to-one on V, further $\varphi(v) = \psi(U)$. Indeed, let U_0, resp. V_0 be a symmetric neighborhood of 0 in $\mathbb{R}^r$, resp. a neighborhood of the unit of G such that we have

$$U_0^2 \cap \mathbb{Z}^r = \{0\}\,, V_0^2 \cap N = \{e\}\,,$$

and $\varphi(V_0) \subseteq \psi(U_0)$. Then we define $U = U_0 \cap \psi^{-1}(\varphi(V_0))$, further $V = V_0 \cap \varphi^{-1}(\psi(U))$. Without loss of generality we may suppose that

$U = \{x \in \mathbb{R}^r : ||x|| < \alpha\}$. For each x in U let $\Phi(x)$ denote the unique element in V such that $\varphi(\Phi(x)) = \psi(x)$ holds, then we have $\Phi = \varphi^{-1} \circ \psi$. It follows that Φ is a continuous local isomorphism of U onto V, that is, if x, y and $x+y$ belong to U, then $\Phi(x+y) = \Phi(x)\Phi(y)$, further $\Phi(-x) = \Phi(x)^{-1}$. We extend Φ to the whole $\mathbb{R}^r$ in the following way: if x is in $\mathbb{R}^r$, then there is a positive integer n such that $y = \frac{x}{n}$ belongs to U, and in this case we define $\Phi(x) = \Phi(y)^n$. It is easy to see that this assignment defines Φ uniquely at x, further the extension Φ is a continuous and open homomorphism of $\mathbb{R}^r$ into G. As $\Phi(\mathbb{R}^r)$ includes some neighborhood of the identity in G, and G is connected, therefore $\Phi(\mathbb{R}^r) = G$. As U generates $\mathbb{R}^r$ and $\varphi \circ \Phi(x) = \psi(x)$ holds for each x in U, we infer $\varphi \circ \Phi = \psi$. Hence the kernel H of the homomorphism Φ is a subset of the kernel $\mathbb{Z}^r$ of the homomorphism ψ. If $H = \{0\}$, then G is topologically isomorphic to $\mathbb{R}^r$. If $H \neq \{0\}$, then there is a basis$e_1, e_2, \ldots, e_r$ in $\mathbb{Z}^r$, and there are positive integers $d_1, d_2, \ldots, d_k$ $(1 \leqslant k \leqslant r)$ such that the elements $d_1e_1, d_2e_2, \ldots, d_ke_k$ generate H. Obviously $e_1, e_2, \ldots, e_r$ is a basis of $\mathbb{R}^r$, hence each coset of H contains an element of the form $x_1e_1 + x_2e_2 + \cdots + x_re_r$, for which we have

$$0 \leqslant x_1 < d_1\,, 0 \leqslant x_2 < d_2\,, \ldots, 0 \leqslant x_k < d_k\,,$$

and $x_{k+1}, \ldots, x_r$ are real numbers, moreover different elements of this type belong to different cosets, therefore $\mathbb{R}^r/H$ is topologically isomorphic to $\mathbb{T}^k \times \mathbb{R}^{r-k}$, which implies

$$G = \mathbb{R}^r/\text{Ker}\ \Phi = \mathbb{R}^r/H = \mathbb{T}^k \times \mathbb{R}^{r-k}\,.$$

Now let G be an arbitrary group satisfying the conditions, that is $G/N = \mathbb{T}^r \times F$, where N is a finitely generated discrete subgroup of G, r is a nonnegative integer, and F is a finite Abelian group. Let φ be the natural homomorphism of G onto G/N and let π be the projection of $\mathbb{T}^r \times F$ onto $\mathbb{T}^r$. Then $\pi \circ \varphi$ is an open and continuous homomorphism of G onto $\mathbb{T}^r$, whose kernel M is the union of a finite number of cosets of N:

$$M = N \cup x_1N \cup \cdots \cup x_lN\,.$$

As N is discrete, hence it is closed, and every x_jN is closed, too. As these sets for each $x_j \neq e$ do not contain the unit element, therefore N is a neighborhood of the unit in M, hence it is closed and open, which implies that M is a discrete and finitely generated subgroup of G. On the other hand, $G/M = \mathbb{T}^r$. If $r = 0$, then $M = G$, hence G is finitely generated, which implies $G = \mathbb{Z}^k \times F$ with some nonnegative integer k, and finite Abelian group F. Now let $r > 0$. As M is discrete, the homomorphism

$\pi \circ \varphi$ is one-to-one, that is, it is a homeomorphism on some neighborhood U of the unit element of G. As $\mathbb{T}^r$ is locally connected, we may suppose that U is connected and symmetric. Then U^n is connected for each natural number n, hence $\bigcup_{n=1}^{\infty} U^n$ is a connected and open subgroup of G, which is obviously closed, too, hence it is the connected component C of the identity element of G. As $\mathbb{T}^r$ is connected, hence $\pi \circ \varphi(U)$ generates $\mathbb{T}^r$, therefore

$$\mathbb{T}^r = \pi \circ \varphi(C) = \pi \circ \varphi(G),$$

thus $T^r = C/C \cap M$. As C is connected, and $C \cap M$ is a discrete and finitely generated subgroup of it, by the first part of the proof, C is an elementary connected group, that is, $C = \mathbb{T}^k \times \mathbb{R}^{r-k}$ holds for some nonnegative integer k.

As $\pi \circ \varphi(C) = \pi \circ \varphi(G)$, we have $G = CM$. By the Homomorphism Theorem

$$G/C = CM/C = M/M \cap C.$$

Actually here we have isomorphism instead of equality, but these isomorphisms are topological isomorphisms at the same time, as the groups in question are discrete, hence we can identify them. Consequently, G/C is finitely generated, hence $G/C = \mathbb{Z}^l \times F$ holds for some nonnegative integer l, and finite Abelian group F. As C is open and divisible, hence G is topologically isomorphic to $C \times (G/C)$, and the proof is complete. □

Theorem 8.7. *(Approximation Theorem) In a compactly generated locally compact Abelian group every neighborhood of the identity contains a compact subgroup such that the corresponding factor group is an elementary group.*

Proof. By Theorem 8.2, in the compactly generated locally compact Abelian group G there is a finitely generated discrete subgroup N such that the factor group $G^* = G/N$ is compact. Let f denote the natural homomorphism of G onto G/N, and let V be an arbitrary neighborhood of the identity e in G, finally, let $W \subseteq V$ be a symmetric neighborhood of the identity with compact closure such that $W^3 \cap N = \{e\}$. By the previous theorem, G^* has a closed subgroup $H^* \subseteq f(W)$ such that G^*/H^* is an elementary group. Then we let $H' = f^{-1}(H^*)$ and $H = H' \cap W$. We note that f is a homeomorphism of W^{cl} onto $f(W^{cl})$, hence it is a homeomorphism of H onto H^*. It follows that H is compact. We show that $HH^{-1} \subseteq H$, that is, H is a subgroup. Indeed, if x, y are in H, then

xy^{-1} belongs to H', and there exists a z in H such that $f(z) = f(xy^{-1})$. Therefore $xy^{-1}z^{-1}$ belongs to N and to W, hence $xy^{-1} = z$ is in H. Now we show that $H' = HN$. Let z be arbitrary in H', then there exists an x in H such that $f(x) = f(z)$, hence $y = zx^{-1}$ belongs to N. It follows $H' = HN$. As $H \subseteq W$, hence the only common element of H and N is the identity. The natural image of N in the factor group G/H is denoted by $\widetilde{N}$. The previous argument shows that $\widetilde{N}$ is isomorphic to N, that is, it is a finitely generated discrete group. On the other hand the factor group of G/H with respect to $\widetilde{N}$ is isomorphic to G/H', as

$$(G/H) \,/\, \widetilde{N} = (G/H) \,/\, N = (G/H) \,/\, (HN/H) = G/HN = G/H' ,$$

that is, $(G/H) \,/\, \widetilde{N}$ is isomorphic to G^*/H^*, which is a compact elementary group. Hence G/H is an elementary group. □

8.3 Duality theory of locally compact Abelian groups

Now we can prove Pontryagin's Duality Theorem first in the compactly generated case, and then in full generality.

Theorem 8.8. *(Pontryagin) Compactly generated locally compact Abelian group is topologically isomorphic to $\mathbb{R}^n \times \mathbb{Z}^k \times K$, where n, k are nonnegative integers, and K is a compact Abelian group.*

Proof. By Theorem 8.7, G has a compact subgroup H such that the corresponding factor group is an elementary group, that is,

$$G/H = \mathbb{R}^n \times \mathbb{Z}^k \times F ,$$

where F is a finite Abelian group. Let φ be the natural homomorphism of G onto G/H, and let $K = \varphi^{-1}(F)$. Then K is a compact subgroup of G, actually it is the largest compact subgroup of G, as it is easy to see. By the Homomorphism Theorem, it follows

$$G/K = (G/H) \,/\, (K/H) = \mathbb{R}^n \times \mathbb{Z}^k \times F \,/\, F = \mathbb{R}^n \times \mathbb{Z}^k .$$

By Zorn's Lemma 3.2, we have that G has a minimal closed subgroup L such that $G = LK$, in other words, K has a direct complement. Then we have $L \cap K = \{e\}$, hence G is topologically isomorphic to $L \times K$, moreover, L is topologically isomorphic to $G/K = \mathbb{R}^n \times \mathbb{Z}^k$. Finally we have

$$G = L \times K = \mathbb{R}^n \times \mathbb{Z}^k \times K .$$

□

Theorem 8.9. *The natural mapping of a compactly generated locally compact Abelian group onto its second dual is a topological isomorphism.*

Proof. Obvious. □

Theorem 8.10. *Let G be a locally compact Abelian group, and let H be an open subgroup of G. Then the mapping, assigning to each character of G its restriction to H, is an open and continuous homomorphism of $\widehat{G}$ onto $\widehat{H}$, the kernel of which is the annihilator of H. Hence the dual of H is topologically isomorphic to the factor group of the dual of G with respect to the annihilator of H.*

Proof. It is clear that the mapping, assigning to each character of G its restriction to H, is a homomorphism of $\widehat{G}$ into $\widehat{H}$, and every character of an open subgroup can be extended to a character of the whole group, hence this mapping is surjective. The statement about the continuity and openness of this mapping is a simple consequence of the definition of the topologies in question. □

Theorem 8.11. *(Pontryagin's Duality Theorem) The natural mapping of a locally compact Abelian group is a topological isomorphism onto its second dual.*

Proof. Let H be a compactly generated open subgroup of G, and let Φ denote the natural mapping of G into its second dual. Clearly $\Phi(x)$ belongs to $A(\widehat{\widehat{G}}, A(\widehat{G}, H))$, the annihilator of $A(\widehat{G}, H)$, for each x in H. We show that each element f of $A(\widehat{\widehat{G}}, A(\widehat{G}, H))$ is of the form $\Phi(x)$ with some x in H. If χ_1 and χ_2 are characters of G such that they are identical on H, then $f(\chi_1) = f(\chi_2)$. Let ψ be a character of H, then, by the previous theorem, ψ is the restriction of some character χ of G onto H. Let $f_0(\psi) = f(\chi)$. By the previous consideration, this definition is independent of the choice of χ, and f_0 maps the dual of H into $\mathbb{T}$. Obviously f_0 is a homomorphism, which is also continuous, as by the previous theorem, the mapping assigning to each character its restriction is open. It follows that f_0 is a character of $\widehat{H}$, by the version of the duality theorem, formulated in Theorem 8.9, holds for H, hence $f_0(\psi) = \psi(x)$ is fulfilled for each ψ in $\widehat{G}$ with some element x in H, that is $f(\psi) = \psi(x) = \chi(x)$ holds for each x in H and χ in $\widehat{G}$. This means that Φ maps H onto the annihilator of $A(\widehat{G}, H)$, obviously isomorphically. As H is open, hence G/H is discrete, and the dual of G/H is $A(\widehat{G}, H)$, which implies that $A(\widehat{G}, H)$ is compact. The annihilator of a compact subgroup is always open, hence $\Phi(H) = A(\widehat{\widehat{G}}, A(\widehat{G}, H))$ is open, therefore Φ is open and continuous, that is, it is a topological isomorphism of H onto $\Phi(H)$, hence of G onto $\Phi(G)$.

Finally, we have to prove that $\Phi(G) = \widehat{\widehat{G}}$. We show that for each f in $\widehat{\widehat{G}}$ there is a compactly generated open subgroup H of G such that f belongs to $\Phi(H)$. Let $U(F,\varepsilon)$ be a neighborhood of the identity of $\widehat{G}$ such that $|f(\chi) - 1| < \sqrt{3}$, whenever χ is in $U(F,\varepsilon)$. Let U be symmetric neighborhood of the identity of G with compact closure, and let H_1 denote the subgroup generated by U. Then H_1 is an open and compactly generated subgroup, hence G/H_1 is discrete. Let ρ be the topological isomorphism of $(G/H_1)\hat{}$ onto $A(\widehat{G}, H_1)$ defined by the following equation

$$\rho(\psi) = \psi \circ \varphi\,, (\psi \in (G/H_1)\hat{})\,,$$

where φ is the natural homomorphism of G onto G/H_1. As

$$U(F,\varepsilon) \cap A(\widehat{G}, H_1)$$

is a neighborhood of the identity in $A(\widehat{G}, H_1)$, therefore

$$\rho^{-1}(U(F,\varepsilon) \cap A(\widehat{G}, H_1))$$

is a neighborhood of the identity in $(G/H_1)\hat{}$, and there exists a compact, hence finite subset $x_1H, x_2H, \ldots, x_nH$ in G/H_1 such that it generates a subgroup C with the property

$$A((G/H_1)\hat{}, C) \subseteq \rho^{-1}(U(F,\varepsilon) \cap A(\widehat{G}, H_1))\,.$$

Now let H be an open and compactly generated subgroup of G such that it contains the closure of U, and the elements $x_1, x_2, \ldots, x_n$. If χ is in $A(\widehat{G}, H)$, then χ takes the value 1 at the elements of H_1, and $\chi(x_j) = 1$ for each $j = 1, 2, \ldots, n$. Thus χ belongs to $A(\widehat{G}, H_1)$, and $\rho^{-1}(\chi)$ belongs to $A((G/H_1)\hat{}, C)$. Hence χ is in $U(F,\varepsilon)$, which implies

$$|f(\chi) - 1| < \sqrt{3}\,.$$

Consequently, $|f(\chi) - 1| < \sqrt{3}$ holds for each χ in $A(\widehat{G}, H)$, and $A(\widehat{G}, H)$ is compact, hence we have $f(\chi) = 1$ for each χ in $A(\widehat{G}, H)$. It follows that f is in $A(\widehat{\widehat{G}}, A(\widehat{G}, H))$, therefore it is in $\Phi(H)$, too. □

Chapter 9

HAAR INTEGRAL ON LOCALLY COMPACT ABELIAN GROUPS

9.1 Haar measure and Haar integral

We have seen in Chapter 6.2 that the Haar measure and the Haar integral on compact Abelian groups serve as basic tools for Fourier analysis and abstract harmonic analysis. In the previous Chapter 7.7 Pontryagin's Duality Theorem has been extended to arbitrary locally compact Abelian groups. Now we want to extend the fundamental results of the theory of Fourier transformation to this more general situation, too. For this purpose we need a corresponding Haar measure on non-compact locally compact Abelian groups. It seems to be reasonable to study if there exists a measure on every locally compact Abelian groups, which possesses the properties listed in Theorem 7.4. It turns out that this is possible in the non-compact case only if we omit the first property. We formulate the following definitions.

Let X be a locally compact Hausdorff space. A measure defined on the σ-algebra of all Borel subsets of X is called *regular measure*, if

1. the measure of each open set is equal to the supremum of the measures of compact subsets of it, and
2. the measure of each Borel set is equal to the infimum of the measures of open sets containing it.

Let G be a locally compact Abelian group. A nonzero regular measure μ defined on the σ-algebra of all Borel subsets of G is called a *Haar measure* on G, if

1. the measure μ is *translation invariant*: for each Borel set B and element g in G we have $\mu(g \cdot B) = \mu(B)$, and

2. every compact set has finite measure.

It is clear that for compact G, apart from a positive constant factor, there exists a unique measure on G with these properties, and it is a positive constant multiple of the one introduced in Chapter 6.2. It is also obvious that there exists such a Haar measure on every discrete Abelian group, as any positive constant multiple of the counting measure possesses the listed properties. On the other hand, we have seen that on compact Abelian groups the existence of Haar measure is equivalent to the existence of a translation invariant positive linear functional on the space of continuous functions. On locally compact Hausdorff spaces the Riesz Representation Theorem 7.1 sets up a link between measures and linear functionals. This motivates us to formulate the following definition.

Let G be a locally compact Abelian group. The nonzero positive linear functional F on the space $\mathcal{C}_0(G)$ is called *Haar integral* on G, if for each f in $\mathcal{C}_0(G)$ and g in G we have

$$F(f) = F(\tau_g f)\,.$$

We have the following theorem.

Theorem 9.1. *Let G be a locally compact Abelian group. If μ is a Haar measure on G then the functional F defined for each f in $\mathcal{C}_0(G)$ by*

$$F(f) = \int f\,d\mu \tag{9.1}$$

is a Haar integral on G. Conversely, if F is a Haar integral on G, then the measure μ corresponding to it by Theorem 7.1, satisfying $F = F_\mu$, is a Haar measure on G.

By the results in Section 8.3 we obtain the following result.

Theorem 9.2. *On a compactly generated locally compact Abelian group there exists a Haar measure, which is unique up to a positive constant multiple.*

Proof. Using the representation of compactly generated locally compact Abelian groups presented in Theorem 8.8, and the obvious fact that our statement is true for each direct factor there, the product of the Haar integrals on the direct factors is exactly the desired Haar integral. □

By this theorem we get a Haar measure on arbitrary compactly generated locally compact Abelian groups, which is unique up to a positive constant factor. The existence of Haar measure on arbitrary locally compact Abelian groups will be proved in the following section.

9.2 The existence of Haar measure on locally compact Abelian groups

The results in this section are taken from [Izzo (1992)]. We start with some technical lemmas.

Lemma 9.2.1. *Let G be a group, and let A, B, C be arbitrary subsets of G. Then we have*

$$(AB) \cap C \neq \varnothing \qquad \text{if and only if} \qquad B \cap (A^{-1}C) \neq \varnothing . \tag{9.2}$$

Proof. The set $(AB) \cap C$ is nonempty if and only if

$$e \in (AB)^{-1}C = B^{-1}(A^{-1}C),$$

and this holds if and only if $B \cap (A^{-1}C)$ is nonempty. □

Lemma 9.2.2. *Let G be a locally compact Abelian group, and let U be a symmetric neighborhood of the identity. Then there exists a subset S in G such that for each element a in G the set aU^2 contains at least one element of S, and the set aU contains at most one element of S.*

Proof. Let $\mathcal{K}$ denote the family of all subsets T of G for which

$$q \notin pU^2$$

holds whenever $p \neq q$ are in T. We note that in this case p cannot be in qU^2, neither. Indeed, if $p = qu_1u_2$ holds for some elements $u - 1, u_2$ in U, then $q = pu_2^{-1}u_1^{-1}$, which belongs to pU^2, by the symmetry of U.

It is obvious that each subset of G including at most one element belongs to $\mathcal{K}$, and it is easy to see that $\mathcal{K}$ satisfies the conditions of Zorn's Lemma 3.2, hence it has a maximal element S. Let a be arbitrary in G, and suppose that aU^2 does not contain any element of S, that is

$$S \cap aU^2 = \varnothing .$$

Now we apply Lemma 9.2.1 with the choice $A = U^2$, $B = \{a\}$ and $C = S$, then, by the symmetry of U, we have

$$\{a\} \cap U^2S = \varnothing ,$$

that is a is not in U^2S. It follows that a does not belong to sU^2 for each s in S. This implies that s does not belong to aU^2, hence the set $S \cup \{a\}$ is in $\mathcal{K}$, which contradicts the maximality of S. This means that for each element a in G the set aU^2 includes at least one element of S.

Suppose now that there are elements $s_1 \neq s_2$ in S such that

$$s_1 \in aU\,, \qquad \text{and} \qquad s_2 \in aU\,.$$

Then $\{s_1\} \cap (aU)$ is nonempty, and we can apply Lemma 9.2.1 with the choice $A = A^{-1} = U$, $B = \{s_1\}$ and $C = \{a\}$. We infer

$$s_1 U \cap \{a\} \neq \varnothing\,,$$

hence a is in $s_1 U$, and a^{-1} is in $s_1^{-1} U$. Similarly, we conclude that a is in $s_2 U$. Consequently

$$e \in s_1^{-1} s_2 U^2\,.$$

This means that the set $s_1^{-1} s_2 U^2$ is nonempty. Now we apply Lemma 9.2.1 with the choice $A = \{s_1\}$, $B = \{e\}$ and $C = s_2 U^2$, and we get

$$s_1 \in s_2 U^2\,,$$

which contradicts the definition of S. Hence the set aU contains at most one element of S, and the theorem is proved. □

Theorem 9.3. *There exists a Haar integral on every locally compact Abelian group.*

Proof. The proof is based on the Markov–Kakutani Fixed Point Theorem 4.9. Let the locally compact Abelian group G be given. We denote by $\mathcal{C}_c(G)^*$ the dual space of $\mathcal{C}_c(G)$ with the weak*-topology. For each a in G we denote, as earlier, by $T_a : \mathcal{C}_c(G)^* \to \mathcal{C}_c(G)^*$ the mapping defined by

$$T_a\,\Lambda(f) = \Lambda(\tau_a f)$$

for each f in $\mathcal{C}_c(K)$. Then each T_a is a continuous linear operator of the locally convex topological vector space $\mathcal{C}_c(G)^*$. Obviously, all operators T_a commute. To prove our statement we have to show that there is a nonzero positive linear functional on $\mathcal{C}_c(G)$, which is a fixed point of each operator T_a. For this it is enough to present a nonempty compact convex subset in $\mathcal{C}_c(G)^*$, which is mapped into itself by all operators T_a, because in this case our statement is a consequence of the Markov–Kakutani Fixed Point Theorem 4.9. First of all we choose a compact symmetric neighborhood U of the identity in G. Let C be the set of all positive linear functionals Λ, which possess the following two properties:

1. $\Lambda(f) \leqslant 1$, whenever f is in $\mathcal{C}_c(G)$ with $0 \leqslant f \leqslant 1$, further the support of f is included in aU for some a in G;

2. $\Lambda(f) \geqslant 1$, whenever f is in $\mathcal{C}_c(G)$ with $0 \leqslant f$, further $f = 1$ on the set aU^2 for some a in G.

It is clear that C is a closed convex subset of $\mathcal{C}_c(G)^*$. By Theorem 3.17 on the Partition of Unity we can see that every nonnegative element of $\mathcal{C}_c(G)$ is a finite sum of nonnegative continuous functions, each of them has its support in a set aU with some a in G. Hence, by the first condition in the definition of C, it follows that for each function f in $\mathcal{C}_c(G)$ the set $\{\Lambda(f) : \Lambda \in C\}$ is bounded. Then, by repeating the argument used in the proof of Tikhonov's Theorem 3.4, we get that the set C is compact. Obviously, all operators T_a map C into itself. Finally, we have to show only that C is nonempty. However, by choosing the set S in Lemma 9.2.2 for this neighborhood U, then the linear functional $f \mapsto \sum_{s\in S} f(s)$ belongs to C.

We conclude that, by the Markov–Kakutani Fixed Point Theorem 4.9, all operators T_a have a common fixed point, which is exactly a Haar integral on G, and our theorem is proved. □

9.3 The uniqueness of the Haar integral

In this section we prove that on locally compact Abelian groups the Haar integral is unique up to a positive constant factor. Although we have proved this in Theorem 9.2 in the compactly generated locally compact situation, but now we give an independent proof for the more general case. We shall use the following fundamental theorem. Different versions of this theorem can be found in [Halmos (1950); Loomis (1953); Hewitt and Ross (1979)].

Theorem 9.4. *(Fubini) Let $(X, \mathcal{A}, \mu)$ and $(Y, \mathcal{B}, \nu)$ be complete measure spaces, and $f : X \times Y \to \mathbb{C}$ a $\mu \times \nu$-integrable function. Then the iterated integrals*

$$\int_X \Big(\int_Y f(x,y)\, d\nu(y)\Big)\, d\mu(x), \qquad \int_Y \Big(\int_X f(x,y)\, d\mu(x)\Big)\, d\nu(y)$$

exist, and

$$\int_{X\times Y} f\, d(\mu\times\nu) = \int_X \Big(\int_Y f(x,y)\, d\nu(y)\Big)\, d\mu(x) = \int_Y \Big(\int_X f(x,y)\, d\mu(x)\Big)\, d\nu(y).$$

Theorem 9.5. *Given two Haar integrals on a locally compact Abelian group any of them is a positive constant multiple of the other.*

Proof. Let I and J be two Haar integrals on G, f a compactly supported continuous function, and h a nonzero compactly supported continuous symmetric function, that is, $h(x) = h(x^{-1})$ holds for each x in G. By Fubini's[1] Theorem 9.4, and by the invariance of the two integrals we can proceed as follows:

$$I(h)J(f) = I_y J_x h(y) f(x) = I_y J_x h(y) f(xy)\,,$$

and

$$J(h)I(f) = I_y J_x h(x) f(y) = I_y j_x h(y^{-1}x) f(y)$$

$$= J_x I_y h(x^{-1}y) f(y) = J_x I_y h(y) f(xy)\,.$$

This implies $I(h)J(f) = I(f)J(h)$, and, by $I(h) \neq 0$, our statement follows immediately. □

9.4 Convolution

To introduce convolution we use the following result.

Theorem 9.6. *(Generalized Young's[2] Inequality) Let $(X, \mathcal{A}, \mu)$ be a measure space, $1 \leqslant p \leqslant +\infty$ and $C > 0$ real numbers. If $K : X \times X \to \mathbb{C}$ is a measurable function for which*

$$\int |K(x,y)|d\mu(y) \leqslant C \qquad \textit{for almost all } x \textit{ in } X\,,$$

and

$$\int |K(x,y)|d\mu(x) \leqslant C \qquad \textit{for almost all } y \textit{ in } X\,,$$

then for each f in $L^p(X)$ the function Tf defined by

$$Tf(x) = \int K(x,y) f(y) d\mu(y)$$

for each x in X belongs to $L^p(X)$, and we have

$$\|Tf\|_p \leqslant C\|f\|_p\,.$$

[1] Guido Fubini, Italian mathematician (1879-1943)
[2] William Henry Young, English mathematician (1863-1942)

Proof. The statement is obvious if $p = +\infty$.

Let $1 \leqslant p < +\infty$, and let q be defined by $\frac{1}{p} + \frac{1}{q} = 1$. By Hölder's Inequality 7.12, we have

$$|Tf(x)| \leqslant \left(\int |K(x,y)|d\mu(y)\right)^{\frac{1}{q}} \left(\int |K(x,y)||f(y)|^p d\mu(y)\right)^{\frac{1}{p}}$$

$$\leqslant C^{\frac{1}{q}} \left(\int |K(x,y)||f(y)|^p d\mu(y)\right)^{\frac{1}{p}}$$

whenever x is in X. Taking p-th powers on both sides and integrating with respect to x, by Fubini's Theorem 9.4, we have

$$\int |Tf(x)|^p d\mu(x) \leqslant C^{\frac{p}{q}} \int\int |K(x,y)||f(y)|^p d\mu(y) d\mu(x)$$

$$\leqslant C^{\frac{p}{q}+1} \int |f(y)|^p d\mu(y)\,.$$

Taking p-th roots we infer

$$\|Tf\|_p \leqslant C^{\frac{1}{p}+\frac{1}{q}} \|f\|_p = C \cdot \|f\|_p\,.$$

Hence $Tf(x)$ is finite for almost all x in X, and our theorem is proved. □

W. H. Young

O. L. Hölder

G. Fubini

Let G be a locally compact Abelian group. We say that the function $f : G \to \mathbb{C}$ is *locally integrable*, if it is measurable, and its restriction to every compact subset is integrable. For instance, every continuous function on G is locally integrable. Let f, g be locally integrable functions on G. If for some x in G the function

$$y \mapsto f(x-y)g(y)$$

is integrable, then we define the *convolution* of f and g by the formula

$$f * g(x) = \int f(x-y)g(y)\, dm_n(y)$$

at the point x. Obviously $f * g(x) = g * f(x)$.

Theorem 9.7. *(Young's Inequality) Let $1 \leqslant p \leqslant +\infty$. If f is in $L^1(G)$ and g is in $L^p(G)$, then for almost all x in G there exists $f * g(x)$, moreover the function $f * g : x \mapsto f * g(x)$ belongs to $L^p(G)$, and*

$$\|f * g\|_p \leqslant \|f\|_1 \cdot \|g\|_p .$$

Proof. The statement follows from the Generalized Young's Inequality 9.6 with the choice $K(x, y) = f(x - y)$. □

The statement of this theorem can be symbolically expressed in the form $L^1 * L^p \subseteq L^p$. In particular, we have $L^1 * L^1 \subseteq L^1$, that is, the space L^1 is closed under convolution.

Theorem 9.8. *The space $L^1(G)$, equipped with the linear operations, with the convolution, and with the norm $\|,\|_1$, is a commutative Banach algebra.*

Proof. The associativity is a consequence of Fubini's Theorem 9.4. □

By Hölder's Inequality 7.12, if $1 \leqslant p \leqslant +\infty$, and q is such that

$$\frac{1}{p} + \frac{1}{q} = 1 ,$$

then for each f in $L^p(G)$ and g in $L^q(G)$ the function $f * g$ belongs to $L^\infty(G)$, and we have

$$\|f * g\|_\infty \leqslant \|f\|_p \cdot \|g\|_q . \tag{9.3}$$

Here, as usual, in the case $p = 1$ we mean $q = +\infty$, and if $p = +\infty$, then $q = 1$.

9.5 Haar measure on elementary and on compactly generated Abelian groups

By our above consideration, we can summarize what is important to know about Haar measure and Haar integral on elementary groups, and on compactly generated Abelian groups. First of all we note that it follows from general measure theory that the product of the Haar measures on finitely many locally compact Abelian groups is a Haar measure on their direct sum. Hence the structure theorems of elementary, respectively compactly generated locally compact Abelian groups make it possible to describe the Haar measure on these groups completely. First we consider the case of elementary groups.

By definition, the elementary groups have the form

$$G = \mathbb{R}^n \times \mathbb{Z}^k \times \mathbb{T}^m \times F$$

where n, k, m are nonnegative integers, and F is a finite Abelian group. It is clear that on $\mathbb{R}^n$ the Lebesgue measure, and on $\mathbb{Z}^m$ the counting measure possesses the characterizing properties of the Haar measure, hence on the group G any Haar measure μ is a positive constant multiple of the product of these measures and the measure given on the group $\mathbb{T}^m \times F$ by equation (7.11). If λ denotes the Lebesgue measure on $\mathbb{R}$, σ denotes the counting measure on $\mathbb{Z}$, and $|F|$ denotes the number of elements in the group F, then, in general, we use the following Haar integral on G:

$$\int f\,d\mu = \frac{1}{(2\pi)^{m+\frac{n}{2}}|F|} \sum_{l\in\mathbb{Z}^k} \sum_{v\in F} \int_{\mathbb{R}^n} \int_{[0,2\pi]^m} f(x,l,t,v)\,d\lambda^n(x)\,d\lambda^m(t)\,. \tag{9.4}$$

Obviously every elementary group is compactly generated and locally compact, hence the previous formula is a special case of the corresponding one for compactly generated locally compact Abelian groups. As all these groups have the form

$$G = \mathbb{R}^n \times \mathbb{Z}^k \times K\,,$$

by Theorem 8.8, where K is a compact Abelian group, hence the Haar integral corresponding to the Haar measure μ can be written in the form

$$\int f\,d\mu = \frac{1}{(2\pi)^{\frac{n}{2}}} \sum_{l\in\mathbb{Z}^k} \int_{\mathbb{R}^n} \int_K f(x,l,k)\,d\lambda^n(x)\,d\kappa(k)\,, \tag{9.5}$$

where κ is a Haar measure on K.

Chapter 10

HARMONIC ANALYSIS ON LOCALLY COMPACT ABELIAN GROUPS

10.1 Harmonic analysis on the group of integers

Let $G = \mathbb{Z}$ be the additive group of integers with the discrete topology. As on every discrete Abelian group, a Haar measure on $\mathbb{Z}$ is an arbitrary positive constant multiple of the counting measure. When studying compact Abelian groups we have seen, in connection with the Inversion Theorem, that the constant factor should be chosen to 1, which we shall assume in our forthcoming investigations. Hence, by Haar measure on discrete Abelian groups we always mean the counting measure. The dual of $\mathbb{Z}$ is identified with $\mathbb{T}$, and this identification is given by the mapping $z \longleftrightarrow \chi_z$, where $|z| = 1$ is a complex number, and the character χ_z is defined by

$$\chi_z(n) = z^n,$$

for each integer n. Obviously, this dual group, as we have seen in Section 7.3, can also be identified with the group defined on $[0, 2\pi]$, equipped it with addition modulo 2π, where $0 \leqslant t < 2\pi$. This identification is given by the mapping $e^{it} \longleftrightarrow \chi_t$, where the character χ_t is defined by

$$\chi_t(n) = e^{int},$$

for each integer n.

Hence, the space $L^2(\mathbb{Z})$ is the set of all functions $f : \mathbb{Z} \to \mathbb{C}$, for which the sum

$$\sum_{x \in \mathbb{Z}} |f(x)|^2$$

is finite. This space obviously can be identified with the space $l^2(\mathbb{Z})$.

If we want to keep on moving on the road we followed in the case of compact Abelian groups, then we need to define Fourier transformation.

Although the concept of Haar integral is available, but still we have to face a serious problem: the characters of $\mathbb{Z}$ definitely do not belong to $L^2(\mathbb{Z})$, hence they never form an orthonormal base in it, consequently we cannot use the general theory of Hilbert spaces, even in the case of the simplest infinite discrete Abelian group. What is the way out of this virtually hopeless situation? We obviously know that, for instance, on the real line the classical Fourier transformation works excellently, as an indispensable tool of modern analysis and its applications. How to include it in the abstract theory we are going to build up?

In what follows we shall see that the solution is provided by the theory of Banach algebras, more exactly, the theory of L^1-algebras. Moreover, there is no need to restrict this theory to the case of the group $\mathbb{Z}$: using the results in Chapter 7.7 we can start immediately discussing harmonic analysis on general locally compact Abelian groups. Nevertheless, a detailed treatment of the theory of Banach algebras is out of the scope of this volume, hence in the subsequent section we shall give a general survey on these necessary, however very deep results. For the proofs and other details see [Loomis (1953)].

10.2 Commutative algebras

Let G be a locally compact Abelian group, additively written, and m will denote an arbitrary Haar measure on it. It is known that the $L^1(G)$ space corresponding to m is a Banach space, if equipped with pointwise addition and multiplication with scalars, further with the norm defined by

$$\|f\| = \int |f|\, dm\,. \tag{10.1}$$

Moreover, as we have seen in Theorem 9.8, equipped with convolution this space is a commutative Banach algebra. Still we can move one more step forward: if for each f in $L^1(G)$ and x in G the function $f^* : G \to \mathbb{C}$ is defined by

$$f^*(x) = \overline{f(x^{-1})}\,, \tag{10.2}$$

and is called the *involution* of f, then $L^1(G)$ is a commutative B^*-algebra, which is called the L^1*-algebra* of the group G. As it is easy to see, $L^1(G)$ has a unit exactly if G is discrete, and in this case the unit element of this algebra is the characteristic function of the singleton consisting of the identity of G. In the case, when G is non-discrete, we have to face some difficulties due to the lack of a unit in $L^1(G)$, because we cannot apply the theory of commutative unital B^*-algebras directly. In the subsequent paragraphs it will be outlined how to overcome this difficulty.

Let A be a commutative algebra, and let I be an ideal in A. We say that I is a *maximal ideal*, if it is not a proper subset of any proper ideal. We say that the element u of the algebra A is an *identity modulo* I, if for each x in A the element $ux - x$ belongs to I. The ideal I is called *regular*, if there exists an identity modulo I in A. If I is a proper regular ideal, then it does not contain any identity modulo I.

Theorem 10.1. *Every proper regular ideal of the commutative algebra A is included in some regular maximal ideal.*

Proof. Let I be a proper regular ideal, and let u be an identity modulo I. All ideals of A containing I and not including u form a partially ordered set with respect to set-theoretical inclusion, which – as it is easy to see – satisfies the condition of Zorn's Lemma 3.2. It follows that there is a maximal element in this set, which – as it is easy to show – is a regular maximal ideal containing I. □

In a commutative algebra the element y is called the *adverse* of the element x, if $x + y - xy = 0$ holds. It is easy to see that the element x has no adverse if and only if x is an identity modulo some regular maximal ideal.

The *spectrum* of an element x in a commutative algebra with identity is called the set of all complex numbers λ for which the element $x - \lambda e$ has no inverse, e being the identity of the algebra. In a commutative algebra without identity the spectrum of the element x is the set of all complex numbers λ for which $\lambda = 0$, or the element x/λ has an adverse.

Theorem 10.2. *Every regular maximal ideal in a commutative Banach algebra is closed, and the respective factor algebra is isometrically isomorphic to the Banach algebra of complex numbers. Hence every regular maximal ideal is equal to the kernel of some continuous homomorphism onto the Banach algebra of complex numbers. Conversely, the kernel of every such homomorphism is a regular maximal ideal.*

Continuous homomorphisms of a Banach algebra onto the Banach algebra of complex numbers are called *multiplicative functionals.* The mapping $h \longleftrightarrow \operatorname{Ker} h$, which assigns to each multiplicative functional its kernel is a one-to-one mapping between all multiplicative functionals and all regular maximal ideals of the Banach algebra. The set of all multiplicative functionals is called the *structure space* or *maximal ideal space* of the Banach algebra, and we denote it by Δ.

Let x be an arbitrary element of the Banach algebra A. For each h in Δ the function $\widehat{x}$ defined by

$$\widehat{x}(h) = h(x) \tag{10.3}$$

is called the *Gelfand*[1] *transform* of x, while the mapping $x \mapsto \widehat{x}$ is the *Gelfand transformation.* It is easy to see that

$$\|\widehat{x}\|_\infty \leqslant \|x\| \tag{10.4}$$

holds for each x in A. The following theorem is of fundamental importance (see [Loomis (1953)]).

Theorem 10.3. *If A is a commutative Banach algebra, then the topology induced by the family of functions $\widehat{A} = (\widehat{x})_{x\in A}$ on the structure space Δ is locally compact, and it is compact if A has an identity. If Δ is not compact, then the functions in $\widehat{A}$ vanish at infinity.*

Hence the family of functions $\widehat{A}$, that is the range of the Gelfand transformation, is a set of continuous functions on a locally compact Hausdorff space, which is obviously an algebra with respect to the pointwise operations. Nevertheless, it is not true in general that this set is the whole of $\mathcal{C}(\Delta)$ or $\mathcal{C}_0(\Delta)$, and it is not even dense in any of these spaces. However, the fundamental theorem of Gelfand theory is the following.

I. M. Gelfand

Theorem 10.4. *(Gelfand–Naimark*[2] *Theorem) The maximal ideal space of a commutative B^*-algebra with identity is a compact Hausdorff space, and the algebra is isometrically $*$-isomorphic to the B^*-algebra of all continuous complex valued functions on the maximal ideal space. The isomorphism is given by the Gelfand transformation.*

[1] Israil Mojseejevic Gelfand, Russian mathematician (1913-2009)
[2] Mark Aronovich Naimark, Russian mathematician (1909-1978)

This theorem presents a complete description of all commutative B^*-algebras with identity by identifying them to the B^*-algebra of all continuous complex valued functions on a compact Hausdorff space. It has particular importance that this identification is performed by a specific mapping, the Gelfand transformation. We shall see how to apply this theorem in the special case of locally compact Abelian groups for the Fourier transformation of integrable functions.

M. A. Naimark

In the non-unital case we have the following result.

Theorem 10.5. *The maximal ideal space of a commutative B^*-algebra without identity is a locally compact Hausdorff space, and the algebra is isometrically $*$-isomorphic to the B^*-algebra of all continuous complex valued functions vanishing at infinity on the maximal ideal space. The isomorphism is given by the Gelfand transformation.*

10.3 The Fourier transform of integrable functions

Let G be a locally compact Abelian group and we fix a Haar measure m on G. We have seen above that with the convolution

$$f * g(x) = \int f(xy^{-1})g(y)\, dm(y)\,,$$

defined for functions f, g in $L^1(G)$ and for x in G, the space $L^1(G)$ is a commutative algebra, which, equipped with the norm $\|.\|_1$ and with the involution defined by (10.2), is a commutative B^*-algebra. The following theorem is of basic importance.

Theorem 10.6. *Let G be a locally compact Abelian group. For each multiplicative functional Φ of the algebra $L^1(G)$ there exists a unique character χ_Φ of G such that*

$$\Phi(f) = \int f\, \overline{\chi}_\Phi\, dm\,. \tag{10.5}$$

Proof. It is well-known (see e.g. [Halmos (1950)]) that for every measure space $(X, \mathcal{M}, \mu)$ each linear functional φ of the Banach space $L^1(X)$ has a representation in the form

$$\varphi(f) = \int f\,\overline{\alpha}_\varphi\, d\mu\,,$$

where f is an arbitrary element in $L^1(X)$, and α is a unique element of L^∞. As every multiplicative functional Φ of $L^1(G)$ is a linear functional, applying this we get

$$\Phi(f) = \int f\,\overline{\alpha}_\Phi\, dm$$

for each f in $L^1(G)$ with some bounded measurable function $\alpha_\Phi : G \to \mathbb{C}$, which is almost everywhere uniquely determined. Let f be in $L^1(G)$, and let x, y be arbitrary elements in G. Then we can proceed as follows:

$$\Phi(\tau_x f) = \int f(xz)\overline{\alpha_\Phi(z)}\, dm(z) = \int f(z)\overline{\alpha_\Phi(x^{-1}z)}\, dm(z)$$

$$= \alpha_\Phi(x) \int f(z)\overline{\alpha_\Phi(z)}\, dm(z) = \alpha_\Phi(x) \cdot \Phi(f)\,,$$

and putting xy for x and $\tau_y f$ for f we get

$$\alpha_\Phi(xy)\Phi(f) = \Phi(\tau_{xy} f) = \Phi\big(\tau_y(\tau_x f)\big)$$

$$= \alpha_\Phi(y)\Phi(\tau_x f) = \alpha_\Phi(y) \cdot \alpha_\Phi(x) \cdot \Phi(f)\,.$$

It follows

$$\alpha_\Phi(xy) = \alpha_\Phi(x) \cdot \alpha_\Phi(y)\,,$$

as Φ is not the zero functional. As α_Φ is measurable, it is known from measure theory that α_Φ is almost everywhere equal to a continuous function. Finally, as α_Φ is bounded, its absolute value is necessarily 1, hence it is a character. □

From this theorem it follows that the maximal ideal space Δ of $L^1(G)$, as a set, can be identified with the dual of G. It is easy to check that the topology of Δ is identical with the dual topology on $\widehat{G}$, the compact-open topology introduced in Section 5.5. Consequently, in what follows, in the case of the $L^1(G)$ it is reasonable to identify the maximal ideal space Δ with $\widehat{G}$. This identification makes it possible to identify the Fourier transformation on $L^1(G)$ with the Gelfand transformation on $L^1(G)$, as a

B^*-algebra. Hence for each function f in $L^1(G)$ and character χ of G we let

$$\widehat{f}(\chi) = \int f\,\overline{\chi}\,. \tag{10.6}$$

The function $\widehat{f} : \widehat{G} \to \mathbb{C}$ is called the *Fourier transform* of f, and the mapping $f \mapsto \widehat{f}$ is the *Fourier transformation.* It is clear that if G is a compact Abelian group, and f is in $L^2(G)$, then this coincides with the Fourier transformation defined on compact Abelian groups. The following statement is a consequence of the Stone–Weierstrass Theorem 3.57, and the general properties of the Gelfand transformation.

Theorem 10.7. *Let G be a locally compact Abelian group. If G is compact, then the Fourier transformation is a contracting $*$-homomorphism of $L^1(G)$ onto a dense subspace of $\mathcal{C}(G)$. If G is not compact, then the Fourier transformation is a contracting $*$-homomorphism of $L^1(G)$ onto a dense subspace of $\mathcal{C}_0(G)$.*

10.4 Fourier transformation on locally compact Abelian groups

In this section we summarize the most important results on the Fourier transformation on locally compact Abelian groups.

Theorem 10.8. *Let G be a locally compact Abelian group. If f is in $L^1(G)$, x is in G, and χ, η are in $\widehat{G}$, then we have*

1. $(\tau_x f)\widehat{\ }(\chi) = \chi(x)\widehat{f}(\chi)\,,$
2. $(\eta f)\widehat{\ }(\chi) = \widehat{f}(\eta^{-1}\chi)\,.$

Proof. We can compute as follows:

$$(\tau_x f)\widehat{\ }(\chi) = \int f(xy)\overline{\chi(y)}\,dm(y) = \chi(x)\int f(xy)\overline{\chi(xy)}\,dm(y)$$

$$= \chi(x)\int f(y)\overline{\chi(y)}\,dm(y)\,,$$

and

$$(\eta f)\widehat{\ }(\chi) = \int \eta(y)\,f(y)\overline{\chi(y)}\,dm(y) = \int f(y)\overline{\eta^{-1}(y)\,\chi(y)}\,dm(y)$$

$$= \widehat{f}(\eta^{-1}\chi)\,.$$

□

Theorem 10.9. *(Convolution Theorem) Let G be a locally compact Abelian group. Then for each f, g in $L^1(G)$ we have*

$$(f * g)\widehat{\ } = \widehat{f} \cdot \widehat{g}\,. \tag{10.7}$$

Proof. The statement follows from the general properties of the Gelfand transformation. □

Theorem 10.10. *(Uniqueness Theorem) Let G be a locally compact Abelian group. If for the function f in $L^1(G)$ we have $\widehat{f} = 0$, then f vanishes almost everywhere.*

Proof. The statement follows from the general properties of the Gelfand transformation. □

Theorem 10.11. *(Inversion Theorem) Let G be a locally compact Abelian group. Then the Haar measure m on G and the Haar measure $\widehat{m}$ on $\widehat{G}$ can be chosen in the way that if f is in $L^1(G)$ and $\widehat{f}$ is in $L^1(\widehat{G})$, then we have for almost all x in G*

$$f(x) = \int \widehat{f}(\chi)\chi(x)\,d\widehat{m}(\chi)\,. \tag{10.8}$$

If f is continuous, then this holds for each x.

Theorem 10.12. *(Plancherel's Theorem) Let G be a locally compact Abelian group. Then the restriction of the Fourier transformation to $L^1(G) \cap L^2(G)$ can uniquely be extended to a unitary operator from $L^2(G)$ onto $L^2(\widehat{G})$.*

PART 2

Spectral Analysis and Synthesis

Chapter 11

BASIC CONCEPTS

11.1 Basics from ring theory

We shall use basic concepts and facts from commutative ring theory. As reference books we depend on the terminology and notation of some standard volumes like [Atiyah and Macdonald (1969); Nagata (1975); Zariski and Samuel (1975a,b); Matsumura (1980); Jacobson (1985, 1989); Dummit and Foote (2004)]. Here we shortly recall the basic concepts, facts, notation, and terminology we shall use in the sequel.

By a *ring* we always mean a commutative ring with identity. In any ring R we shall use the standard notation $A + B$ and $A \cdot B$ for the *ideal sum* and the *ideal product*, resp. of the ideals A, B, which denote the smallest ideal containing all sums $a + b$, resp. $a \cdot b$ with a in A and b in B. In particular, the *ideal power* is defined recursively, in the natural way with the convention $A^0 = R$.

A proper ideal in a ring is called a *maximal ideal*, if it is not properly contained in any proper ideal. The *nilradical*, or simply *radical* of an ideal is the set of all elements having a positive power belonging to the ideal. The radical of I is denoted by $\operatorname{Rad} I$. An ideal in a ring is called *prime*, if it contains the product of two elements if and only if at least one of them belongs to the ideal. Every maximal ideal is prime. An ideal is called *primary*, if it contains a product of two elements the one of which not belonging to the ideal, then the other belongs to the radical of the ideal. Clearly, each prime ideal is primary, and the radical of each primary ideal is prime. Further, each ideal is primary, if its radical is maximal. A ring is called *local ring*, if it has exactly one maximal ideal. A ring is called *semi-local*, if it has finitely many maximal ideals.

If R is a ring, H is a subset in R, and A is an R-module, then we use the notation $H \cdot A$, or HA for the set of all elements in A of the form $h \cdot a$ with h in H and a in A.

The ideals I, J in the ring R are called *co-prime*, if their sum is R: $I + J = R$. A family of ideals in R is called *pairwise co-prime*, if any two ideals in the family are co-prime. We have the following simple result.

Theorem 11.1. *In the ring R let $I_1, I_2, \ldots, I_k$ be pairwise co-prime ideals. Then we have*

$$I_1 \cap I_2 \cap \cdots \cap I_k = I_1 \cdot I_2 \cdot \cdots \cdot I_k \,.$$

Proof. For $k = 2$ we have that $e = x_1 + x_2$ with some x_1 in I_1 and x_2 in I_2, where e is the identity in R. Clearly, $I_1 \cdot I_2 \subseteq I_1 \cap I_2$. On the other hand, let x be in $I_1 \cap I_2$. Then $x \cdot x_1$ is in $I_1 \cdot I_2$, as x_1 is in I_1 and x is in I_2, further $x \cdot x_2$ is in $I_1 \cdot I_2$, as x is in I_1 and x_2 is in I_2. Hence $x = x \cdot x_1 + x \cdot x_2$ is in $I_1 \cdot I_2$, and $I_1 \cap I_2 \subseteq I_1 \cdot I_2$.

For $k \geqslant 3$ we observe that $I_1 + I_2 = R$, and $I_2 + I_3 = R$, hence it follows $I_1 \cdot I_3 + I_2 \cdot I_3 = I_3$. Then we have

$$I_1 \cdot I_2 + I_3 = I_1 \cdot I_2 + (I_1 + I_2) \cdot I_3 = I_1 \cdot I_2 + I_1 \cdot I_3 + I_2 \cdot I_3$$

$$= I_1 \cdot I_2 + I_1 \cdot I_3 + I_1 \cdot I_3 + I_2 \cdot I_3 = I_1 \cdot (I_2 + I_3) + (I_1 + I_2) \cdot I_3 = I_1 + I_3 = R\,,$$

that is, $I_1 \cdot I_2$ and I_3 are co-prime, and our statement follows immediately, by induction. □

Lemma 11.1.1. *Any powers of different maximal ideals are co-prime.*

Proof. Let M_1, M_2 be different maximal ideals in the ring R, further let k, n be natural numbers. Suppose that $M_1^{k+1} + M_2^{n+1} \neq R$, then there exists a maximal ideal M such that $M_1^{k+1} + M_2^{n+1} \subseteq M$. It follows $M_1^{k+1} \subseteq M$ and $M_2^{n+1} \subseteq M$. As M is prime, we infer $M_1, M_2 \subseteq M$, which implies, by maximality, $M_1 = M_2 = M$, a contradiction. □

Another useful lemma is the following.

Lemma 11.1.2. *(Chinese Remainder Theorem) Let R be a ring, and let $I_1, I_2, \ldots, I_n$ be pairwise co-prime ideals in R. Then R/I has a direct decomposition in the form*

$$R/I \cong R/I_1 \oplus R/I_2 \oplus \cdots \oplus R/I_n\,.$$

Conversely, if R/I has a direct decomposition of the form

$$R/I \cong R_1 \oplus R_2 \oplus \cdots \oplus R_n$$

with some rings $R_1, R_2, \ldots, R_n$, then there are pairwise co-prime ideals $I_1, I_2, \ldots, I_n$ in R such that $I = I_1 \cap I_2 \cap \cdots \cap I_n$ and $R_k \cong R/I_k$ for $k = 1, 2, \ldots, n$.

Proof. It is easy to check that the homomorphism

$$F : R \to R/I_1 \oplus R/I_2 \oplus \cdots \oplus R/I_n$$

given by

$$F(x) = (x + I_1, x + I_2, \ldots, x + I_n)$$

is surjective, and its kernel is I. The converse is obvious. □

A ring is called *Noether ring*[1], if it satisfies the *ascending chain condition* for the ideals: any ascending chain of ideals terminates after finitely many steps. In a Noether ring every ideal is finitely generated. A ring is called an *Artin ring*[2], if it satisfies the *descending chain condition* for the ideals: any descending chain of ideals terminates after finitely many steps. Every Artin ring is Noether ring, but the converse is not true in general. Analogously, a module is called Noether module, resp. Artin module, if it satisfies the ascending, resp. descending chain condition for the submodules. It is not true in general, that Artin modules are Noether modules.

E. Noether

E. Artin

W. Krull

11.2 Vector modules

Let R be a ring, and let X be a topological vector space. We say that X is a *vector module over R* if X, as an Abelian group, is a module

[1]Emmy Noether, German mathematician (1882-1935)
[2]Emil Artin, Austrian mathematician (1898-1962)

over R, and for each r in R the mapping $x \mapsto r \cdot x$ is a *continuous linear operator* on X, further the linear operator $x \mapsto e \cdot x$ is the *identity operator* on X. We remark that if no topology is specified on X, then we always consider it with the *discrete topology*. By a *vector submodule*, or simply a *submodule*, of a vector module we mean a linear subspace, which is also a vector module over R, with the same meaning of $r \cdot x$, of course. A closed vector submodule is called a *variety*. It is obvious that the intersection of any nonempty family of submodules, resp. varieties is a submodule, resp. variety. For any subset H, resp. element x in X the smallest submodule, resp. variety is the intersection of all submodules, resp. varieties including H, resp. x, which is called the *submodule*, resp. *variety generated by H, resp. x*.

If X is a topological vector space, then $\mathcal{L}(X)$ denotes the algebra of all continuous linear mappings, that is, the linear operators on X. On $\mathcal{L}(X)$ one usually considers the *strong operator topology*, hence all topological concepts on this space refer to that topology. In this topology a generalized sequence $(A_i)_{i\in I}$ of operators converges to the operator A if and only if the generalized sequence $\big(A_i(x)\big)_{i\in I}$ converges to $A(x)$ in X for each x in X.

Let X be a topological vector space, and let R be a ring. By a *representation* of the ring R on X we mean a homomorphism of R into $\mathcal{L}(X)$. If R is a *topological ring*, and this homomorphism is continuous, then we call it a *continuous representation*. If R has an identity, then we require that it is mapped onto the identity operator. Similarly, if an algebra $\mathcal{A}$ is given, then a *representation* of this algebra on X we mean a representation of the ring on X, which is also a homomorphism of the linear space structure of $\mathcal{A}$. Continuity of an algebra representation is meant in the obvious way.

The following theorem clears up the connection between vector modules and representations.

Theorem 11.2. *Let X be a topological vector space, and suppose that a representation of the ring R is given on X. If at this representation the element r of R is mapped onto the operator A_r in $\mathcal{L}(X)$, then, by defining $r \cdot x = A_r x$ for each r in R and x in X, X is a vector module over R. Every vector module uniquely arises in this way.*

Proof. Let $\Phi : R \to \mathcal{L}(X)$ denote the given representation, that is,

$$\Phi(r) = A_r$$

holds for each r in R. Then we have for each x, y in X and r, s in R

$$r \cdot (x + y) = A_r(x + y) = A_r x + A_r y = r \cdot x + r \cdot y\,,$$

$$(r+s)\cdot x = A_{r+s}x = \Phi(r+s)(x) = \Phi(r)(x)+\Phi(s)(x) = A_r x+A_s x = r\cdot x+s\cdot x\,,$$

$$(rs) \cdot x = A_{rs}x = \big(A_r A_s\big)x = A_r\big(A_s x\big) = r \cdot (s \cdot x)\,,$$

hence X, as an Abelian group, is a module over R. As A_r is continuous, X is a vector module.

Suppose now that X is a vector module over the ring R. Then, by definition, the mapping $A_r : x \mapsto r \cdot x$ is a continuous linear mapping, that is, a linear operator on X, and clearly the mapping $\Phi : r \mapsto A_r$ is a representation of R on X, which induces the given vector module structure on X.

The uniqueness statement is obvious. □

This theorem shows that any vector module can be realized as an ordered pair consisting of a topological vector space X, and a ring of linear operators on X. This is similar to the situation of ordinary modules: these are pairs consisting of an Abelian group, and a ring of endomorphisms of it. In the case of vector modules the submodules are exactly those subspaces, which are invariant under the operators belonging to the given ring. But these are exactly those linear subspaces of X, which are invariant under the linear operators belonging to the operator algebra generated by the ring of linear operators in question. Hence, in what follows, we may always suppose that our vector modules over rings are actually vector modules over operator algebras of linear operators on the given topological vector space. More exactly, we shall consider vector modules over operator algebras, by which we mean a topological vector space X together with a unital algebra $\mathcal{A}$ of linear operators on X. Then the submodules of this vector module are exactly the $\mathcal{A}$-invariant subspaces of X. Here we give some simple examples.

1. Let X be a vector space over the field F. Then the scalar operators in $\mathcal{L}(X)$, that means, the scalar multiples of the identity operator, form an operator algebra $\mathcal{A}$ on X, and X is a vector module over $\mathcal{A}$. Submodules are exactly the linear subspaces of X.

2. Let X be a topological vector space and let A be a linear operator on X. Then X is a vector module over the operator algebra $\mathcal{A}_A$ generated by A. We always suppose, unless the contrary is explicitly stated, that this algebra includes the identity operator, too. Submodules are exactly the A-invariant linear subspaces of X.
3. Let $X = \mathbb{C}^{\mathbb{N}}$ denote the vector space of all complex sequences with the product topology, and let τ be the *shift operator* defined by

$$(\tau x)_n = x_{n+1}$$

for each sequence $(x_n)_{n\in\mathbb{N}}$ in X and n in N. Then X is a vector module over the operator algebra $\mathcal{A}_\tau$ generated by the shift operator. Submodules are exactly the shift invariant linear spaces of sequences.

11.3 Vector modules, group representations, and actions

Let X be a topological vector space and G a group. By a *representation* of G on X we mean a homomorphism of G into $\mathcal{L}(X)$, where the unit element of G is mapped onto the identity operator. If G a topological group, and the homomorphism is continuous, then we call it a *continuous representation.* If T denotes this homomorphism, then let $\mathcal{A}_T$ denote the subalgebra in $\mathcal{L}(X)$ generated by the image of G under T. Obviously, $\mathcal{A}_T$ is the set of all finite linear combinations of operators of the form $T(g)$ with g is in G. Then X is a vector module over the operator algebra $\mathcal{A}_T$. We say that this vector module is *induced by the representation* T. Submodules are those linear subspaces, which are invariant under all operators $T(g)$ with g in G. We may call them *G-invariant subspaces*, however these depend not just on G, but rather on T. We remark, that if T is a representation of G on X, then we may write T_g instead of $T(g)$.

Let E be a topological space, and suppose that a topological group G is given, which acts continuously on E. This means, that a continuous map $\pi : G \times E \to E$ is given with $\pi\big(g_1, \pi(g_2, a)\big) = \pi(g_1 g_2, a)$, and $\pi(e, a) = a$ for each a in E and g_1, g_2 in G, where e denotes the identity of G. The map π will be referred to as an *action* of G, and the function $x \mapsto \pi(g, x)$ will be denoted by π_g for each g in G.

Let X be an arbitrary topological vector space of complex valued functions on E. Suppose that this function space is *π-invariant*, which means that for each f in X and g in G the function $T_\pi(g)f$, defined by

$$T_\pi(g)f(a) = f\big(\pi(g,a)\big) \tag{11.1}$$

whenever a is in E, belongs to X. Suppose, moreover, that $f \mapsto T_\pi(g)f$ is a linear operator on X. Then $T_\pi : g \mapsto T_\pi(g)$ is a representation of G on X. In this case we say that this representation is *induced by the action* π.

This means that if a continuous action of G on E is given, which induces the representation T_π of G on the π-invariant function space X, then X becomes a vector module over the algebra $\mathcal{A}_{T_\pi}$. The vector submodules of X are exactly those linear subspaces of X, which are π-invariant. If an action π of G on E is given, and X is a π-invariant function space on E, then we may call X a *π-module.*

11.4 Spectral analysis and synthesis on vector modules

Let X be a vector module over the algebra $\mathcal{A}$. We say that *$\mathcal{A}$-spectral analysis*, or simply *spectral analysis* holds on X, if every nonzero subvariety of X has a nonzero finite dimensional submodule, or, what is the same, a nonzero finite dimensional subvariety. By definition, spectral analysis holds for the zero variety, and spectral analysis holds for a variety if and only if it holds for each of its subvarieties. We say that X is *$\mathcal{A}$-synthesisable*, or simply *synthesisable*, if the sum of all finite dimensional vector submodules of X is dense in X. We say that *$\mathcal{A}$-spectral synthesis*, or simply *spectral synthesis* holds for X, if each subvariety of X is $\mathcal{A}$-synthesisable. Again, by definition, spectral synthesis holds for the zero variety, and spectral synthesis holds for a variety if and only if it holds for each of its subvarieties. Observe that spectral analysis means that *there are* nonzero finite dimensional varieties in every nonzero subvariety, and on the other hand, spectral synthesis means that *there are sufficiently many* nonzero finite dimensional varieties in every nonzero subvariety. Clearly, spectral synthesis implies spectral analysis, however, we shall see that – apart from trivial cases – the converse fails to hold. If $\mathcal{A}$ is of the form $\mathcal{A}_A$, or $\mathcal{A}_T$, resp. $\mathcal{A}_{T_\pi}$, defined as above, then we speak about A-spectral analysis and A-spectral synthesis, or T-spectral analysis and T-spectral synthesis, resp. π-spectral analysis and π-spectral synthesis. Here we give some simple examples.

1. Obviously, spectral synthesis holds for each finite dimensional vector module.

2. Let X be a topological vector space over a field F. Then, as we have seen, X is a vector module over the algebra of scalar operators, and the submodules are exactly the linear subspaces of X. It follows that spectral synthesis holds for X, as the sum of all finite dimensional subspaces in every closed subspace of X is dense.
3. This example shows the connection between spectral analysis and the invariant subspace problem (see e.g. [Abramovich *et al.* (2005)], [Yadav (2005)]). Let X be a topological vector space, and let A be a linear operator in $\mathcal{L}(X)$. As we have seen, X is a vector module over the algebra $\mathcal{A}_A$, and the subvarieties are exactly the closed A-invariant linear subspaces of X. Hence A-spectral analysis for X is equivalent to the existence of a nonzero finite dimensional invariant subspace of A in each nonzero closed A-invariant subspace of X.
4. Suppose that X is a Banach space, and A is a compact operator on X. By the spectral theory of compact operators, X is the sum of eigensubspaces of A, and the same holds for each A-invariant closed subspace. Moreover, each eigensubspace corresponding to a nonzero element of the spectrum of A is finite dimensional. It follows that A-spectral analysis holds for each A-invariant closed subspace, which has a finite dimensional intersection with the kernel of A. Moreover, A-spectral synthesis holds for X if and only if the kernel of A is finite dimensional.

The following theorem is important.

Theorem 11.3. *Let X be a vector module over the algebra $\mathcal{A}$. Spectral analysis holds on X if and only if given any nonzero subvariety X_0 of X there is a positive integer n, and there are linearly independent vectors $x_1, x_2, \ldots, x_n$ in X_0 such that*

$$Ax_i = \sum_{j=1}^{n} \lambda_{i,j}(A)x_j \tag{11.2}$$

holds for $i = 1, 2, \ldots, n$ and for each A in $\mathcal{A}$ with some functions $\lambda_{i,j} : \mathcal{A} \to \mathbb{C}$ satisfying the system of functional equations

$$\lambda_{i,j}(AB) = \sum_{k=1}^{n} \lambda_{i,k}(A)\lambda_{k,j}(B) \tag{11.3}$$

for each A, B in $\mathcal{A}$ and for $i, j = 1, 2, \ldots, n$.

Proof. Supposing that spectral analysis holds on X let X_0 be a nonzero subvariety of X, and let $\{x_1, x_2, \ldots, x_n\}$ be a basis of a nonzero finite dimensional submodule of X_0. Then for each $i, j = 1, 2, \ldots, n$ there exist complex numbers $\lambda_{i,j}$ such that (11.2) holds for $i = 1, 2, \ldots, n$ and for each A in $\mathcal{A}$. Putting AB for A in (11.2) we get

$$ABx_i = \sum_{j=1}^{n} \lambda_{i,j}(AB)x_j\,. \tag{11.4}$$

On the other hand, we have

$$A(Bx_i) = \sum_{k=1}^{n} \lambda_{i,k}(A)(Bx_k) = \sum_{j=1}^{n}\sum_{k=1}^{n} \lambda_{k,j}(B)\lambda_{i,k}(A)x_j\,. \tag{11.5}$$

By the linear independence of the x_j's we get

$$\lambda_{i,j}(AB) = \sum_{k=1}^{n} \lambda_{i,k}(A)\lambda_{k,j}(B)\,, \tag{11.6}$$

and the necessity of the condition of the theorem is proved.

Conversely, suppose that the condition of the theorem is satisfied with some elements $x_1, x_2, \ldots, x_n$ in the nonzero subvariety X_0 of X, and with some complex numbers $\lambda_{i,j}$ in $\mathbb{C}$ $(i, j = 1, 2, \ldots, n)$. Then, by (11.2), the subspace in X_0 spanned by $x_1, x_2, \ldots, x_n$ is nonzero, and it is $\mathcal{A}$-invariant, hence spectral analysis holds on X. □

The functional equation in (11.3) is called *Levi–Cività equation*[3]. The functions $\lambda_{i,j}$ define a representation Λ of the algebra $\mathcal{A}$ on the Hilbert space $\mathbb{C}^n$ in the following way: let $\{e_1, e_2, \ldots, e_n\}$ denote the standard orthonormal basis in $\mathbb{C}^n$, where the j-th component of e_i is $\delta_{i,j}$, the Kronecker's[4] symbol, and we use the Euclidean inner product: $\langle e_i, e_j\rangle = \delta_{i,j}$ for $i, j = 1, 2, \ldots, n$. Then, for each A in $\mathcal{A}$ we let

$$\langle \Lambda(A)e_i, e_j\rangle = \lambda_{i,j}\,,$$

whenever $i, j = 1, 2, \ldots, n$. By (11.3), it follows

$$\Lambda(AB) = \Lambda(A)\Lambda(B)$$

for each A, B in $\mathcal{A}$, which proves our statement. Obviously, $\Lambda(A)$ can be realized as an $n \times n$ matrix. If $\mathcal{A}$ is unital with identity E, then $\Lambda(E)$ is the identity matrix I, moreover, if $\mathcal{A}$ is closed under taking inverses, then for each invertible operator A in $\mathcal{A}$ the matrix $\Lambda(A)$ is regular:

$$I = \Lambda(E) = \Lambda(AA^{-1}) = \Lambda(A)\Lambda(A^{-1})\,,$$

[3]Tullio Levi–Cività, Italian mathematician (1873-1941)
[4]Leopold Kronecker, German mathematician (1823-1891)

which implies

$$\Lambda(A)^{-1} = \Lambda(A^{-1}).$$

Hence, in this case $\Lambda : \mathcal{A} \to GL_n(\mathbb{C})$ is a representation of the group of invertible operators in $\mathcal{A}$.

T. Levi–Cività

L. Kronecker

According to our remarks and to the terminology in [Shulman (2011)], vectors satisfying a system of equations of the form (11.2) are called *matrix elements*. More exactly, an element x in X is called a *matrix element*, if it is contained in a finite dimensional submodule. Clearly, an element of a vector module is a matrix element if and only if it generates a finite dimensional variety. Using this terminology we can formulate the following theorem.

Theorem 11.4. *Let X be a vector module over the algebra $\mathcal{A}$. Spectral analysis holds on X if and only if every nonzero subvariety of X contains a nonzero matrix element. X is synthesisable if and only if all matrix elements span a dense subspace in X.*

In the case of commutative $\mathcal{A}$ we have the following theorem.

Theorem 11.5. *Let X be a vector module over the commutative algebra $\mathcal{A}$. Spectral analysis holds on X if and only if there exists a common eigenvector for $\mathcal{A}$ in every nonzero subvariety of X.*

Proof. Clearly, if $x \neq 0$ is a common eigenvector for all operators in $\mathcal{A}$ in the subvariety X_0, then the one dimensional subspace spanned by x is an invariant subspace in X_0 for all operators in $\mathcal{A}$, hence it is a one dimensional subvariety of X_0. Conversely, if spectral analysis holds on X, and V is a nonzero finite dimensional submodule of X_0, then, by commutativity, there exists a common eigenvector of all operators of $\mathcal{A}$ in V, that is, in X_0. □

Let X be a topological vector module over the algebra $\mathcal{A}$. The family $(X_i)_{i\in I}$ of submodules of X is called a *decomposition* of X, if the algebraic sum of the X_i's is dense in X, however, the algebraic sum of any proper subfamily is not dense in X. Here *algebraic sum* means the set of all finite sums, where the terms are taken from some X_i's. Clearly, if the family $(X_i)_{i\in I}$ is a decomposition of X, then so is the family $(X_i^{cl})_{i\in I}$. This implies that we can always suppose that the members of a decomposition are closed, that is, they are varieties. We shall always suppose this, when speaking about decompositions. It follows that a decomposition has at least two different nonzero members. We say that X is *decomposable*, if it has a decomposition. It is obvious that X is decomposable if and only if it is the closure of the sum of two proper subvarieties. If X is not decomposable, then we call it *indecomposable*. A matrix element is called *indecomposable*, if it generates an indecomposable variety, which is – by definition – finite dimensional.

Theorem 11.6. *Let X be a vector module over the algebra $\mathcal{A}$. Every finite dimensional variety can be expressed as a sum of finitely many indecomposable varieties. Every matrix element can be expressed as a sum of finitely many indecomposable matrix elements.*

Proof. If X_0 is an indecomposable variety, then we are ready. If not, then X_0 is the closure of the sum of two subvarieties $A, B \subseteq X_0$, which are different form X_0. If both are indecomposable, then the proof is finished. Otherwise we can repeat this argument, and we can decompose either A or B, or both into proper subvarieties. Continuing this process the dimensions decrease, hence after finitely many steps we arrive at a finite decomposition of X_0 into indecomposable submodules, and the proof of the first statement is complete. Obviously, the second statement is just a reformulation of the first one. □

As a consequence we get the following theorem.

Theorem 11.7. *Let X be a vector module over the algebra $\mathcal{A}$. Spectral analysis holds on X if and only if in every nonzero subvariety of X there is a nonzero finite dimensional indecomposable variety, or, equivalently, there is a nonzero indecomposable matrix element. Spectral synthesis holds on X if and only if in each subvariety of V the indecomposable matrix elements span a dense subspace.*

Concerning spectral analysis and synthesis with respect to group actions we use the concept of matrix elements, as well. Obviously, we have the corresponding Theorems 11.4 and 11.5, too.

Here we give a simple example for spectral analysis and synthesis with respect to a group action. Let $E \neq \{0\}$ be a complex vector space, and let X denote the set of all complex valued functions on E, equipped with the topology of pointwise convergence and with the linear operations. Then X is a locally convex topological vector space. Let G be the multiplicative group $\{1, -1\}$. We define the action π of G on E by $\pi(\epsilon, x) = \epsilon x$ for ϵ in G and x in E. Obviously, X is π-invariant. By (11.1), we have $T_\pi(1)f = f$ and $T_\pi(-1)f(a) = f(-a) = \check{f}(a)$ for each a in E. It follows that a subspace of X is π-invariant if and only if it contains $\check{f}$ for each f in X_0. On the other hand, common eigenfunctions of $T_\pi(1)$ and $T_\pi(-1)$ are exactly the nonzero even functions: $\check{f} = f$, and the nonzero odd functions: $\check{f} = -f$. Hence spectral analysis holds for X if and only if each nonzero subvariety X_0 in X contains either a nonzero even, or a nonzero odd function. But if X_0 is not zero, then for any nonzero f in X_0 we have that the function $\check{f}$, consequently, also the even function $2f_e = f + \check{f}$ and the odd function $2f_o = f - \check{f}$ belongs to X_0. As $f = f_e + f_o$, it follows that either f_e, or f_o is nonzero, hence spectral analysis holds for each nonzero subvariety. Finally, by $f = f_e + f_o$, we have that actually spectral synthesis holds for each subvariety, too.

11.5 Varieties on groups

Now we consider a very important special case of the general situation exhibited above. Let G be a *locally compact topological group.* In Section 5.5 we introduced the compact-open topology on the *dual group* of G. It turns out that this topology can be utilized on more general function spaces. On the linear space $\mathcal{C}(G)$ of all continuous complex valued functions on G we introduce the *compact-open topology* in the following way: for each compact set K in G and for each f in $\mathcal{C}(G)$ we let

$$p_K(f) = \sup_{x \in K} |f(x)|\,. \tag{11.7}$$

It is easy to see that p_K is a seminorm on $\mathcal{C}(G)$. The topology on the space $\mathcal{C}(G)$, which is induced by the family of seminorms p_K, where K ranges over all compact subsets of G is called the *compact-open topology*, or *topology of compact convergence.* We know from Section 3.9 that a topology induced by a family of seminorms on a vector space is always a *vector topology.* We also have the following theorem.

Theorem 11.8. *The compact-open topology on $\mathcal{C}(G)$ is a Hausdorff vector topology.*

Proof. Let $f \neq 0$ be a function in $\mathcal{C}(G)$, then there exists a point x_0 in G and a neighborhood U of the identity in G with compact closure such that $|f| \geqslant k > 0$ holds on $K = x_0 U^{cl}$, with some positive number k. Then $p_K(f) > 0$, hence the family of seminorms (p_K) is separating, and our statement follows from Theorem 3.38. □

It is obvious that this topology, when restricted to the set of all characters of G, coincides with the topology of the dual group. This means that $\widehat{G}$, as a topological space, can be identified with a closed subspace of $\mathcal{C}(G)$: the identity mapping embeds $\widehat{G}$ homeomorphically into $\mathcal{C}(G)$.

We can see immediately that in the space $\mathcal{C}(G)$ the convergence of a generalized sequence is equivalent to the *uniform convergence* on all compact subsets of G – this motivates the alternative name of the topology: the *topology of compact convergence.* In particular, in the special case, when G is a discrete Abelian group, $\mathcal{C}(G)$ is the set of all complex valued functions on G. As a set, it can be identified with the product space $\mathbb{C}^G$, and the product topology is exactly the *topology of pointwise convergence.* However, the topology of pointwise convergence on $\mathcal{C}(G)$ is clearly identical with the topology of compact convergence, as in the discrete topology of G compact sets are just the finite sets, and uniform convergence on a finite set is the same as pointwise convergence on this set. We note that in the general case $\mathcal{C}(G)$ equipped with the linear operations and with the pointwise multiplication of functions is a complex algebra, and this multiplication, as a two-place function, is continuous, which can be expressed by saying that $\mathcal{C}(G)$ is a *topological algebra.*

We shall consider the following, very important actions of G onto $\mathcal{C}(G)$. Let for each y in G and f in $\mathcal{C}(G)$ the mapping $\pi_l : G \times \mathcal{C}(G) \to \mathcal{C}(G)$ be defined by

$$\pi_l(y, f)(x) = f(yx),$$

whenever x is in G. The mapping $f \mapsto \pi(y, f)$ will be denoted by λ_y and it will be called the *left translation* of G. Similarly, we define the *right translation* ρ_y of G using the action $\pi_r : G \times \mathcal{C}(G) \to \mathcal{C}(G)$ defined by

$$\pi_r(y, f)(x) = f(xy^{-1}),$$

whenever x is in G. In the commutative case in Section 1.1 we called the right translation simply "translation by y", and we denoted it by τ_y. We shall keep on using this terminology and notation in the sequel, if G is Abelian. According to these actions from the left and right we shall use expressions like *left invariant* and *right invariant*, with the obvious meaning.

Using the above terminology, a *left variety* is a closed left invariant submodule of $\mathcal{C}(G)$ with respect to the algebra of linear operators generated by all left translations λ_y with y in G. The *right variety* has similar meaning. Although in the commutative case the left and right translations corresponding to a fixed element are different, however, invariance with respect to all left translations or to all right translations are obviously equivalent. Hence in that case we can refer closed submodules simply as *varieties*: they are closed linear subspaces of $\mathcal{C}(G)$ invariant under all translations. We use the terms *left synthesizable*, *right synthesizable*, *synthesizable*, *left spectral analysis*, *left spectral synthesis*, *right spectral analysis*, *right spectral synthesis*, and *spectral analysis*, *spectral synthesis* in the obvious sense. In case of locally compact groups, when using this terminology we mean it in general with respect to the given actions of G over $\mathcal{C}(G)$, that is, we shall consider $\mathcal{C}(G)$ as a module over the algebra of linear operators generated by the corresponding translation operators.

Whenever H is a subset in $\mathcal{C}(G)$, then $\tau(H)$ denotes the intersection of all varieties containing H, which is obviously a variety, and it is called the *variety generated by* H. In particular, if $H = \{f\}$, a singleton, then we write $\tau(f)$ for H, and we call it the *variety of* f.

11.6 Annihilators

Let X be a vector module over the operator algebra $\mathcal{A}$, and let H be a subset in X. The *annihilator* of H is the set $H^\perp$ of all elements r in $\mathcal{A}$ satisfying $r \cdot x = 0$ for each x in H. The following statement is straightforward.

Theorem 11.9. *Let X be a vector module over the operator algebra $\mathcal{A}$, and let H be a subset in X. Then the annihilator of H is an ideal in $\mathcal{A}$.*

Proof. Clearly, the annihilator of H is closed under addition. On the other hand, for each r, q in $\mathcal{A}$, if r annihilates H, then for each x in H we have $(q \cdot r)x = q(r \cdot x) = q \cdot 0 = 0$, hence $q \cdot r$ annihilates H. Similarly,

$(r \cdot q)x = r(q \cdot x) = 0$, as $q \cdot x$ is in H, and r is in the annihilator of H. □

Theorem 11.10. *Let X be a vector module over the operator algebra $\mathcal{A}$ and let K be a subset in $\mathcal{A}$. Further let*

$$K^{\perp} = \{x : r \cdot x = 0 \text{ for all } r \text{ in } K\}.$$

Then $K^{\perp}$ is a submodule in X.

Proof. Clearly, $x + y$ and λy belongs to $K^{\perp}$, whenever x, y are in $I^{\perp}$, and λ is a complex number. Moreover, for each x in $K^{\perp}$, q in K, and r in $\mathcal{A}$ we have that $q \cdot r$ is in K, hence $q \cdot (r \cdot x) = (q \cdot r) \cdot x = 0$. It follows that $r \cdot x$ is in $K^{\perp}$, hence $K^{\perp}$ is a submodule. □

The submodule $K^{\perp}$ is called the *annihilator* of the set K. It is easy to see that we have the inclusions $K^{\perp\perp} \supseteq K$ and $H^{\perp\perp} \supseteq H$ for each subset K in $\mathcal{A}$ and H in X. Nevertheless, none of these inclusions can be replaced with equality, in general, even in the case if K is an ideal and H is a submodule. In the most important case of groups exhibited in the previous section, however, we have a special situation. First of all, we have the following result (see e.g. [Hewitt and Ross (1970)], p. 551).

Theorem 11.11. *Let G be a locally compact group. Then the dual space of $\mathcal{C}(G)$ can be identified with the space $\mathcal{M}_c(G)$ of all compactly supported regular Borel measures on G.*

Proof. Let μ be in $\mathcal{M}_c(G)$ with $K = \operatorname{supp}\mu$, and we define

$$\Lambda_{\mu}(f) = \int_G f \, d\mu$$

for each f in $\mathcal{C}(G)$. Then we have

$$|\Lambda(f)| \leqslant |\mu|(K) \cdot p_K(f),$$

which implies that Λ is a continuous linear functional on $\mathcal{C}(G)$. Conversely, let Λ be in $\mathcal{C}(G)^*$. Then there exists a compact set K in G and a constant k such that

$$|\Lambda(f)| \leqslant k \cdot p_K(f)$$

holds for each f in $\mathcal{C}(G)$. By Tietze's Extension Theorem 3.8, the restrictions to K of the functions in $\mathcal{C}(G)$ form the space $\mathcal{C}(K)$, hence Λ defines an element Λ_K in $\mathcal{C}(K)^*$. By the Riesz Representation Theorem 7.3, there is a regular Borel measure on K, which can be extended to G by putting $\mu^*(B) = \mu(B \cap K)$ with $\Lambda(f) = \int_G f \, d\mu^*$ for each f in $\mathcal{C}(G)$. □

We note that convolution on $\mathcal{M}_c(G)$ is defined in the usual way: for μ, ν in $\mathcal{M}_c(G)$ and f in $\mathcal{C}(G)$ we let

$$(\mu * \nu)(f) = \int_G \int_G f(xy) d\mu(x)\, d\nu(y) .$$

The identity in $\mathcal{M}_c(G)$ is the Dirac measure δ_e supported at the identity e of G: $\delta_e(f) = f(e)$ for each f in $\mathcal{C}(G)$. Equipped with the convolution the commutative unital algebra $\mathcal{M}_c(G)$ will be called the *measure algebra* of G. Then $\mathcal{C}(G)$ is a vector module over $\mathcal{M}_c(G)$ with respect to the convolution defined by

$$(\mu * f)(x) = \int_G f(xy^{-1})\, d\mu(y) ,$$

whenever μ is in $\mathcal{M}_c(G)$ and f is in $\mathcal{C}(G)$. For each y in G the right translation operator ρ_y can be expressed as

$$\rho_y f = \delta_y * f$$

for each f in $\mathcal{C}(G)$, where δ_y is the Dirac measure supported at y. Indeed, for every x in G we have

$$\delta_y * f(x) = \int_G f(xz^{-1})\, d\delta_y(z) = f(xy^{-1}) = \rho_y f(x) .$$

We agree on that speaking about spectral analysis and synthesis on a locally compact group we always mean it with respect to the measure algebra $\mathcal{M}_c(G)$.

Theorem 11.12. *Let G be a locally compact group, and let V be a variety in $\mathcal{C}(G)$. Then $V^{\perp\perp} = V$.*

Proof. We have to show that $V^{\perp\perp} \subseteq V$. Supposing the contrary there is an f in $V^{\perp\perp}$ such that f does not belong to V. It follows $\mu(f) = 0$ for each μ in $V^{\perp}$. We apply the Hahn–Banach Theorem 3.48: there exists a linear functional λ of $\mathcal{C}(G)$, which annihilates V and $\lambda(f) \neq 0$. This means that λ is in $V^{\perp}$ and $\lambda(f) \neq 0$, a contradiction. □

The corresponding theorem for ideals is not true, in general, as the following simple example shows. Consider $G = \mathbb{R}$ with the usual topology, and let I denote the ideal generated by the measures $\mu_n = \delta_0 - \delta_{1/n}$ for $n = 1, 2, \ldots$. If f is in $I^{\perp}$, then f is periodic mod $1/n$ for every n, and thus, by continuity, f must be constant. Therefore $\delta_0 - \delta_\alpha$ is in $I^{\perp\perp}$ for each α in $\mathbb{R}$. However, $\delta_0 - \delta_\alpha$ is not in I if α is irrational. Indeed, for every positive integer N there is a continuous function f such that f is *periodic* mod $1/n$

for each $n \leqslant N$ in $\mathbb{N}$ but f is not periodic mod α. This easily implies that $\delta_0 - \delta_\alpha$ does not belong to the ideal generated by μ_n for n in $\mathbb{N}$. However, if $\delta_0 - \delta_\alpha$ is I, then $\delta_0 - \delta_\alpha$ belongs to an ideal generated by finitely many of the measures μ_n, which is not the case (see [Laczkovich and Székelyhidi (2007)]).

Nevertheless, we have the following theorem (see [Laczkovich and Székelyhidi (2007)]).

Theorem 11.13. *Let G be a discrete Abelian group. Then $I^{\perp\perp} = I$ holds for every ideal I in $\mathcal{M}_c(G)$.*

Proof. We only have to prove $I^{\perp\perp} \subseteq I$. If ν is in $I^{\perp\perp}$, then $\int_G f\,d\nu = 0$ holds for every f in $I^\perp$. Suppose that ν is not in I. Since I is a linear subspace of $\mathcal{M}_c(G)$, and ν is not in I, there is a linear map $L : \mathcal{M}_c(G) \to \mathbb{C}$ such that L vanishes on I and $L(\nu) \neq 0$. Let $f(x) = L(\delta_x)$ for each x in G. Then

$$L(\mu) = \int_G f\,d\mu \tag{11.8}$$

holds for every μ in $\mathcal{M}_c(G)$. Indeed, (11.8) is true for $\mu = \delta_x$ for every x in G, by the definition of f. As G is discrete, every μ in $\mathcal{M}_c(G)$ is a finite linear combination of measures concentrated on singletons, and (11.8) holds, by the linearity of both sides. Now, if μ in I, then $\int_G f\,d\mu = L(\mu) = 0$ by the choice of L, and hence f belongs to $I^\perp$. On the other hand, $\int_G f\,d\nu = L(\nu) \neq 0$, which contradicts that ν is in $I^{\perp\perp}$. □

On arbitrary locally compact group the following natural question arises: what is the necessary and sufficient condition for the equality $I = I^{\perp\perp}$, whenever I is an ideal in $\mathcal{M}_c(G)$? We obviously have the following lemmas.

Lemma 11.6.1. *Let G be a locally compact group, and let I be an ideal in $\mathcal{M}_c(G)$. Then $I^{\perp\perp\perp} = I^\perp$.*

Proof. Let $V = I^\perp$, then V is a variety on G, hence, by Theorem 11.12, we have $V^{\perp\perp} = V$. It follows $I^\perp = V = V^{\perp\perp} = I^{\perp\perp\perp}$. □

Lemma 11.6.2. *Let G be a locally compact group, and let V be a variety on G. Then $V^\perp$ is weak*-closed in $\mathcal{M}_c(G)$.*

Proof. Let $(\mu_\gamma)_{\gamma\in\Gamma}$ be a generalized sequence in $V^\perp$, which converges to μ in the weak*-topology. This means that for each f in V the generalized sequence $\big(\mu_\gamma(f)\big)_{\gamma\in\Gamma}$ converges to $\mu(f)$. As $\mu_\gamma(f)=0$ for each γ in Γ, we infer $\mu(f)=0$, hence μ is in $V^\perp$, by Theorem 3.5. □

Now we can answer our previous question.

Theorem 11.14. *Let G be a locally compact group, and let I be an ideal in $\mathcal{M}_c(G)$. Then we have $I^{\perp\perp}=I$ if and only if I is weak*-closed.*

Proof. By the previous lemma, the annihilator of each variety is weak*-closed, hence $I^{\perp\perp}$, as the annihilator of the variety $I^\perp$, is weak*-closed, which proves the necessity of our condition.

Conversely, suppose that I is weak*-closed, and I is a proper subset of $I^{\perp\perp}$. Let μ be in $I^{\perp\perp}$ such that μ is not in I. As the space $\mathcal{M}_c(G)$ with the weak*-topology is locally convex, hence, by the Hahn–Banach Theorem 3.50, there is a linear functional ξ in $\mathcal{M}_c(G)^*$, which annihilates I and $\xi(\mu)\neq 0$. By Theorem 3.43, it follows that there is an f in $\mathcal{C}(G)$ with $\xi(\nu)=\nu(f)$ for each ν in $\mathcal{M}_c(G)$. We infer $\mu(f)=\xi(\mu)\neq 0$, and μ is in $I^{\perp\perp}$, hence f is not in $I^{\perp\perp\perp}$. On the other hand, $\nu(f)=\xi(\nu)=0$ for each ν in I, as ξ annihilates I, which implies that f is in $I^\perp$, however, this contradicts to Lemma 11.6.1. Our theorem is proved. □

Theorem 11.15. *Let G be a locally compact group.*

1. *For each family $(V_\gamma)_{\gamma\in\Gamma}$ of varieties on G we have*
$$\Big(\sum_{\gamma\in\Gamma} V_\gamma\Big)^\perp=\bigcap_{\gamma\in\Gamma} V_\gamma^\perp\,.$$
2. *For each family $(I_\gamma)_{\gamma\in\Gamma}$ of ideals in $\mathbb{C}G$ we have*
$$\Big(\sum_{\gamma\in\Gamma} I_\gamma\Big)^\perp=\bigcap_{\gamma\in\Gamma} I_\gamma^\perp\,.$$

We note that here $\sum_{\gamma\in\Gamma} V_\gamma$ denotes the *topological sum* of the family of varieties $(V_\gamma)_{\gamma\in\Gamma}$, that is, the closure of the union of the sums of finite subfamilies. However, $\sum_{\gamma\in\Gamma} I_\gamma$ denotes the *algebraic sum* of the family of ideals $(I_\gamma)_{\gamma\in\Gamma}$, that is, the ideal generated by the sums of finite subfamilies.

Proof. If μ is in $\bigcap_{\gamma\in\Gamma} V_\gamma^\perp$, then μ annihilates each of the varieties V_γ, hence it annihilates every finite sum of these varieties, and, by continuity, μ annihilates the closure of the sums of finite subfamilies. Hence μ annihilates $(\sum_{\gamma\in\Gamma} V_\gamma)^\perp$.

Conversely, if μ annihilates $(\sum_{\gamma\in\Gamma} V_\gamma)^\perp$, then μ annihilates all subvarieties of it, hence it belongs to each $V_\gamma^\perp$. This proves the first statement.

To prove the second statement we take f in $(\sum_{\gamma\in\Gamma} I_\gamma)^\perp$. Then f is annihilated by the sum of any finite subfamily of $(I_\gamma)_{\gamma\in\Gamma}$, in particular, it is annihilated by any of these ideals. Hence it belongs to $I_\gamma^\perp$ for each γ.

For the reverse inclusion we take an f, which is annihilated by each ideal I_γ. Then clearly, f is annihilated by any measure in the ideal generated by finite sums of these ideals, hence f is in $(\sum_{\gamma\in\Gamma} I_\gamma)^\perp$. □

Chapter 12

BASIC FUNCTION CLASSES

12.1 Exponentials

In this chapter we consider commutative groups only, and the group operation will be written as addition. In particular, the identity will be denoted by 0, and $\check{f}$ denotes the function defined by $\check{f}(x) = f(-x)$, the *inversion* of f. If a non-discrete topology is given on the basic group, we always suppose that it is locally compact, however, most concepts we intend to introduce do have a reasonable meaning on any topological group. First we introduce some function classes, which are fundamental from the point of view of spectral analysis and synthesis.

We have seen in the first part of this book that the basic building blocks of classical harmonic analysis are the characters. However, in those cases the functions we intended to analyze were subjected to different growth conditions, like boundedness, square integrability or integrability. If we want to study arbitrary continuous functions, then it turns out that characters are not enough. Consequently, when describing translation invariant spaces of continuous functions we have to introduce a wider class of basic functions. In Section 11.4 we introduced the matrix elements, and Theorem 11.4. shows their fundamental importance in spectral analysis and spectral synthesis – in other words, in the description of varieties. In particular, as we have seen in Theorem 11.7, eminent role is played by the indecomposable matrix elements. We shall see that these functions have nice characterizations in the group situation. On the other hand, in more general situations, where the group is replaced by a more general structure, it is very helpful to obtain a more useful description of matrix elements. We start by introducing some basic function classes.

When studying unbounded functions, it seems to be reasonable to replace characters by more general homomorphisms of the underlying group. These are in close connection with the multiplicative functionals of the measure algebra. A linear functional $\Phi : \mathcal{M}_c(G) \to \mathbb{C}$ is called a *multiplicative functional*, if it is an algebra homomorphism of $\mathcal{M}_c(G)$. As $\Phi(\delta_0) = 1$, it follows that the range of every multiplicative functional is $\mathbb{C}$. As a consequence, we have that the *kernel* of a multiplicative functional is a *maximal ideal* of $\mathcal{M}_c(G)$, and $\mathbb{C}G/\mathrm{Ker}\,\Phi \cong \mathbb{C}$. It is not true, in general, that every maximal ideal of the measure algebra is the kernel of some multiplicative functional: maximal ideals of $\mathcal{M}_c(G)$ are exactly the kernels of algebra homomorphisms of $\mathcal{M}_c(G)$ onto extensions of the complex field. Hence, the kernels of multiplicative functionals represent a special class, which will play a basic role in our investigations, and we shall use the term *exponential maximal ideal* for the kernels of multiplicative functionals. We shall apply this terminology concerning the maximal ideals of any commutative unital complex algebra.

Spectral analysis and spectral synthesis is a study of varieties. The simplest nonzero variety is one dimensional. A normed continuous function $m : G \to \mathbb{C}$ on the locally compact Abelian group G is called an *exponential*, if its variety is one dimensional. Here we recall that *normed* means that m takes the value 1 at 0. We have the following characterization of exponentials.

Theorem 12.1. *Let G be a locally compact Abelian group, and let $f : G \to \mathbb{C}$ be a function. Then the following statements are equivalent.*

1. *f is an exponential.*
2. *f is a continuous homomorphism of G into the multiplicative group of nonzero complex numbers.*
3. *f is a common normed continuous eigenfunction of all translation operators.*

Proof. Let f be an exponential. Then f is continuous, and $\tau(f)$ is one dimensional, in particular, f is nonzero, and every translate of f is a constant multiple of f. In other words, there exists a function $\lambda : G \to \mathbb{C}$ such that for each x, y in G we have

$$f(x+y) = \tau_y f(x) = \lambda(y) \cdot f(x)\,. \qquad (12.1)$$

Putting $x = 0$ it follows $f(y) = \lambda(y)$, and from (12.1) we infer that f is a continuous homomorphism of G into the multiplicative group of nonzero complex numbers.

Now suppose that f is a continuous homomorphism of G into the multiplicative group of nonzero complex numbers, that is

$$f(x+y) = f(x) \cdot f(y) \tag{12.2}$$

holds for each x, y in G. Putting $x = y = 0$ in (12.2) we infer that $f(0)$ is 0 or 1. In the first case, putting $y = 0$ in (12.2), it follows $f = 0$, which is excluded. Hence f is normed. On the other hand, (12.2) can be written in the form

$$\tau_y f = f(y) \cdot f$$

for each y in G, hence f is a common normed continuous eigenfunction of all translation operators.

Finally, suppose that f is a common normed continuous eigenfunction of all translation operators. Then there exists a function $\lambda : G \to \mathbb{C}$ such that for each x, y in G we have (12.1). Obviously, λ is continuous. This means that every translate of f is a constant multiple of f, that is, $\tau(f)$ is at most one dimensional. As f is normed, hence $\tau(f)$ is one dimensional and f is an exponential. □

Theorem 12.2. *Let G be a locally compact Abelian group. For each multiplicative functional Φ of the measure algebra $\mathcal{M}_c(G)$ there exists a unique exponential m such that*

$$\Phi(\mu) = \mu(\check{m}) \tag{12.3}$$

holds for each μ in $\mathcal{M}_c(G)$. Conversely, for each exponential m the function $\Phi : \mathcal{M}_c(G) \to \mathbb{C}$ defined by (12.3) is a multiplicative functional of $\mathcal{M}_c(G)$.

Proof. Let $\Phi : \mathcal{M}_c(G) \to \mathbb{C}$ be a multiplicative functional of $\mathcal{M}_c(G)$. In particular, as Φ is a linear functional of the topological vector space $\mathcal{M}_c(G) = \mathcal{C}(G)^*$, hence, by Theorem 3.43., we have (12.3) for each μ in $\mathcal{M}_c(G)$ with some function m in $\mathcal{C}(G)$. Let x be in G, and we apply (12.3) with $\mu = \delta_{-x}$ to obtain

$$\Phi(\delta_{-x}) = \delta_{-x}(\check{m}) = m(x) \tag{12.4}$$

for each x in G. Then we have

$$m(x+y) = \Phi(\delta_{-x-y}) = \Phi(\delta_{-x} * \delta_{-y}) = \Phi(\delta_{-x})\Phi(\delta_{-y}) = m(x)m(y),$$

and $m(0) = \Phi(\delta_0) = 1$, hence m is an exponential. This computation also shows that m is unique. Conversely, it is easy to see that Φ defined by (12.3) is a multiplicative functional of the measure algebra, whenever m is an exponential. □

This theorem sets up a one-to-one correspondence between multiplicative functionals and exponentials. We note that exponentials sometimes are called *generalized characters* (see e.g. [Bhatt and Dedania (2005)]). The set of all exponential on G will be denoted by G^*. It is obviously an Abelian group with respect to pointwise multiplication. Moreover, as a subspace of $\mathcal{C}(G)$, equipped with the compact-open topology it is closed. However, it is not necessarily locally compact, as it is shown in [Bhatt and Dedania (2005)]. Nevertheless, the mapping $m \mapsto \Phi_m$, where Φ_m is the unique multiplicative functional corresponding to m by (12.3) has nice properties.

Theorem 12.3. *Let G be a locally compact Abelian group. The mapping $m \mapsto \Phi_m$, which assigns to each exponential m the multiplicative functional of $\mathcal{M}_c(G)$ given by (12.3) is a continuous bijection of G^* onto the set of all multiplicative functionals equipped with the weak*-topology.*

Proof. In the previous theorem we have seen that the mapping $m \mapsto \Phi_m$ is a bijection between G^* and the set of all multiplicative functionals. On the other hand, let $(m_i)_{i\in I}$ be a generalized sequence in G^* converging to the exponential m, and let μ be a measure with support K. Clearly, the generalized sequence $(\check{m}_i)_{i\in I}$ converges to $\check{m}$. Then for each $\varepsilon > 0$ there exists an i_0 in I such that for $i \geqslant i_0$ we have

$$p_K(\check{m}_i - \check{m}) < \varepsilon\,.$$

It follows

$$|\Phi_{m_i}(\mu) - \Phi_m(\mu)| = |\mu(\check{m}_i) - \mu(\check{m})| =$$

$$\Big|\int_K (\check{m}_i - \check{m})\,d\mu\Big| \leqslant \int_K |\check{m}_i - \check{m}|\,d\mu \leqslant p_K(\check{m}_i - \check{m})\mu(K) < \varepsilon \cdot \mu(K)\,,$$

which proves our statement. □

For each μ in $\mathcal{M}_c(G)$ and m in G^* we define

$$\widehat{\mu}(m) = \Phi_m(\mu) = \mu(\check{m})\,.$$

The mapping $\widehat{\mu} : G^* \to \mathbb{C}$ is an analogue of the Fourier transform. Indeed, obviously G^* contains $\widehat{G}$, the dual of G, and the restriction of $\widehat{\mu}$ to $\widehat{G}$ is the *Fourier–Stieltjes transform* [1] of μ (see e.g. [Rudin (1990)], Section 1.3.3., p. 15.). The mapping $\mu \mapsto \widehat{\mu}$ is an algebra homomorphism of $\mathcal{M}_c(G)$ onto an algebra of continuous functions defined on G^*. More exactly, we have the following theorem.

[1] Thomas Jan Stieltjes, Dutch mathematician (1856-1894)

Theorem 12.4. *Let G be a locally compact Abelian group. Then the mapping $\widehat{\mu}$ is continuous, and it satisfies*

$$(\mu * \nu)\widehat{\ } = \widehat{\mu} \cdot \widehat{\nu}$$

for each μ, ν in $\mathcal{M}_c(G)$. Further, the mapping $\mu \mapsto \widehat{\mu}$ is an injective algebra homomorphism of $\mathcal{M}_c(G)$ into the algebra of continuous complex functions on G^*.

Proof. Suppose that the generalized sequence $(m_i)_{i\in I}$ tends to m in G^*, then the generalized sequence $(\check{m}_i)_{i\in I}$ tends to $\check{m}$. As this convergence is uniform on the compact support of μ, the continuity of $\widehat{\mu}$ follows. For each m in G^* we have

$$(\mu * \nu)\widehat{\ }(m) = (\mu * \nu)(\check{m}) = \int \int \check{m}(x + y)\, d\mu(x)\, d\nu(y)$$

$$\int \check{m}(x)\, d\mu(x) \int \check{m}(y)\, d\nu(x) = \widehat{\mu}(m) \cdot \widehat{\nu}(m)\,.$$

It follows that $\mu \mapsto \widehat{\mu}$ is an algebra homomorphism. Suppose now that $\mu(\check{m}) = 0$ for each m in G^*. Then the restriction of $\widehat{\mu}$ to $\widehat{G}$ vanishes, hence, by 1.3.6. in [Rudin (1990)], p. 17., it follows that $\widehat{\mu} = 0$. □

We note that the function $\widehat{\mu} : G^* \to \mathbb{C}$ defined above is sometimes called the *Fourier–Laplace transform*[2] of μ (see e.g. [Loomis (1953)], [Benedetto (1975)]).

P. S. Laplace

J. Stieltjes

[2]Pierre-Simon Laplace, French mathematician (1749-1827)

12.2 Modified differences

In this section we consider only discrete Abelian groups. The reason is that we shall need Theorem 11.13. We shall use *modified differences*: given a function f on the Abelian group G and an element y in G we define

$$\Delta_{f;y} = \delta_{-y} - f(y)\,\delta_0\,.$$

For the products of modified differences we use the notation

$$\Delta_{f;y_1,y_2,\dots,y_{n+1}} = \Pi_{i=1}^{n+1}\Delta_{f;y_i}\,,$$

for any natural number n and for each $y_1, y_2, \dots, y_{n+1}$ in G. On the right hand side Π is meant as a convolution product.

For each function $f : G \to \mathbb{C}$ the ideal in $\mathbb{C}G$ generated by all modified differences of the form $\Delta_{f:y}$ with y in G is denoted by M_f. It is reasonable to ask whether M_f is proper ideal.

Theorem 12.5. *Let G be an Abelian group, and let $f : G \to \mathbb{C}$ be a function. The ideal M_f is proper if and only if f is an exponential. In this case $M_f = \tau(f)^\perp$ is an exponential maximal ideal.*

Proof. Suppose that M_f is proper, then $M_f^\perp$ is nonzero. Let $\varphi \neq 0$ be a function in $M_f^\perp$. Then $\Delta_{f;y}$ annihilates φ for each y in G, and we have

$$0 = \Delta_{f;y} * \varphi(x) = \varphi(x+y) - f(y)\varphi(x)\,, \tag{12.5}$$

whenever x, y are in G. Putting $x = 0$ we have $\varphi(y) = \varphi(0)f(y)$, in particular $\varphi(0) \neq 0$, hence, by (12.5), f is an exponential.

If f is an exponential, then f is in $M_f^\perp$, hence $M_f^\perp$ is nonzero, and we infer that M_f is proper.

Now let m be an exponential. Clearly, $\tau(m)$ is a one dimensional variety, hence $\tau(m)^\perp$ is a maximal ideal in $\mathbb{C}G$. If φ is in $M_m^\perp$, then $\varphi = \varphi(0)m$, hence φ belongs to $\tau(m)$. It follows $M_m^\perp \subseteq \tau(m)$, that is, $\tau(m)^\perp \subseteq M_m$. By the maximality of $\tau(m)^\perp$, and it follows $M_m = \tau(m)^\perp$, hence M_m is maximal.

We define for each μ in $\mathbb{C}G$

$$F(\mu) = \mu(\check{m})\,.$$

It is easy to check that $F : \mathbb{C}G \to \mathbb{C}$ is a multiplicative functional. For each y in G we have

$$F(\Delta_{m;y}) = \Delta_{m;y}(\check{m}) = (\delta_{-y} - m(y)\delta_0)(\check{m}) = 0\,,$$

hence $\Delta_{m;y}$ is in the kernel of F for each y in G, consequently, we infer that $M_m \subseteq \operatorname{Ker} F$. By the maximality of these ideals it follows $M_m = \operatorname{Ker} F$, and our theorem is proved. □

For each function $f : G \to \mathbb{C}$ the annihilator of the variety of f is called the *annihilator* of f. From the above results we obtain the following characterization of exponentials via their annihilators.

Theorem 12.6. *Let G be a locally compact Abelian group. The function $f : G \to \mathbb{C}$ is an exponential if and only if it is normed, and its annihilator is an exponential maximal ideal.*

We note that given a locally compact Abelian group, then obviously all mappings of the form $f \mapsto \mu * f$, where μ is a measure on G, are continuous linear operators on $\mathcal{C}(G)$. In sections 2.3 and 5.4 we considered convolution operators on $L^2(G)$, where G is a finite discrete, or a compact Abelian group. We introduce this concept on any locally compact Abelian group G in the obvious way: if μ is in $\mathcal{M}_c(G)$, then we let for each f in $\mathcal{C}(G)$

$$A_\mu(f) = \mu * f\,.$$

Then $A_\mu : \mathcal{C}(G) \to \mathcal{C}(G)$ is called the *convolution operator* corresponding to the measure μ. The following theorem is the analogue of Theorem 2.6.

Theorem 12.7. *Let G be a locally compact Abelian group. Then each convolution operator is a continuous linear operator of the topological vector space $\mathcal{C}(G)$. The common normed eigenfunctions of all convolution operators are exactly the exponentials.*

Proof. Obviously, the convolution operators are linear. To prove that $A_\mu * f$ is continuous for each f in $\mathcal{C}(G)$ we take a generalized sequence $(x_i)_{i\in I}$ in G converging to the point x in G. Let K denote the support of μ. We may suppose that x is in K. Obviously, the set $K + (-K)$ is compact, too. We choose an open sets U with compact support such that $K + (-K) \subseteq U$. We may suppose that $U = -U$. There exists an i_0 in I such that x_i is in U, whenever $i \geqslant i_0$. Let $\varepsilon > 0$ be arbitrary. As f is uniformly continuous on U^{cl}, there exists a symmetric neighborhood V of zero such that $|f(u) - f(v)| < \varepsilon$, whenever u, v is in U^{cl} and $u - v$ is in V. Then there exists a j_0 in I such that $i \geqslant j_0$ implies that x_i is in $x + V$. If $i \geqslant i_0$ and $i \geqslant j_0$ and y is in K, then we have that $x_i - y$ and $x - y$ are in U^{cl}, and $(x_i - y) - (x - y) = x_i - x$ is in V. Hence we have

$$|\mu * f(x_i) - \mu * f(x)| \leqslant \int_K |f(x_i - y) - f(x - y)|\, d|\mu|(y) \leqslant \varepsilon|\mu|(K)\,,$$

which implies the continuity of f at x.

To prove the continuity of A_μ suppose that $(f_i)_{i\in I}$ is a generalized sequence in $\mathcal{C}(G)$ converging uniformly to the zero function on each compact subset of G. Let C be a compact set in G. Then $C + (-K)$ is compact, hence for each $\varepsilon > 0$ there is an i_0 in I such that $|f_i(z)| < \varepsilon$, whenever $i \geqslant i_0$ and z is in $C + (-K)$. We have

$$|\mu * f_i(x)| \leqslant \int_K |f_i(x-y)|\, d|\mu|(y) < \varepsilon \cdot |\mu|(K),$$

whenever x is in C, as $x - y$ is in $C + (-K)$ for each y in K. This proves that the generalized sequence $(\mu * f_i)_{i\in I}$ converges uniformly to the zero function on C.

For each μ in $\mathcal{M}_c(G)$ and exponential m we have

$$\mu * m(x) = \int m(x-y)\, d\mu(y) = \mu(\check{m}) \cdot m(x),$$

hence m is a common eigenfunction of the convolution operators and $m(0) = 1$. Conversely, if m is a common eigenfunction of the convolution operators with $m(0) = 1$, and y is in G, then there is a complex number $\lambda(y)$ such that

$$\lambda(y) \cdot m(x) = \delta_{-y} * m(x) = m(x+y)$$

holds for each x in G. With $x = 0$ we have $\lambda = m$, hence m is an exponential. □

12.3 Automorphisms of the measure algebra

In what follows ideals of the measure algebra will play a crucial role in the characterization of functions, which play an important role in spectral analysis and synthesis. It will be useful to study what properties of the ideals and the corresponding residue rings will be preserved under automorphisms of the measure algebra.

Theorem 12.8. *Let G be a locally compact Abelian group. For each exponential m we define*

$$\Psi_m(\mu)(f) = \mu(f \cdot \check{m}), \tag{12.6}$$

whenever μ is in $\mathcal{M}_c(G)$ and f is in $\mathcal{C}(G)$. Then $\Psi_m : \mathcal{M}_c(G) \to \mathcal{M}_c(G)$ is a topological automorphism of the measure algebra.

Proof. Clearly, $\Psi(\mu)$ is a measure. Given Ψ_m as above, and μ, ν in $\mathcal{M}_c(G)$ we have

$$\Psi_m(\mu * \nu)(f) = (\mu * \nu)(f \cdot \check{m}) = \int\int f(x+y)m(-x-y)\,d\mu(x)\,d\nu(y) =$$
$$\int m(-y)\Big[\int f(x+y)m(-x)\,d\mu(x)\Big]\,d\nu(y) =$$
$$\int m(-y)\Big[\int f(x+y)\,d\Psi_m(\mu)(x)\Big]\,d\nu(y) =$$
$$\int\Big[\int f(x+y)\,d\Psi_m(\mu)(x)\Big]\,d\Psi_m(\nu)(y) = \big(\Psi_m(\mu) * \Psi_m(\nu)\big)(f),$$

whenever f is in $\mathcal{M}_c(G)$. As Ψ_m is obviously linear, it follows that it is an algebra homomorphism. If $\Psi_m(\mu) = 0$ for some μ in $\mathcal{M}_c(G)$, then we have

$$0 = \Psi_m(\mu)(f) = \mu(f \cdot \check{m})$$

for each f in $\mathcal{C}(G)$. Taking $f \cdot m$ instead of f we obtain that $\mu = 0$, that is, Ψ_m is injective. For each μ in $\mathcal{M}_c(G)$ we let $\nu(f) = \mu(f \cdot m)$, whenever f is in $\mathcal{M}_c(G)$. Then obviously ν is in $\mathcal{M}_c(G)$ and $\Psi_m(\nu) = \mu$, hence Ψ_m is surjective. It follows that Ψ_m is an automorphism of the measure algebra. The proof of the continuity of the function Ψ_m and of its inverse is a routine calculation. □

The following theorem will be useful in the sequel.

Theorem 12.9. *Let G be an Abelian group, n a natural number, and let m_1, m_2 be exponentials. Then the rings $\mathbb{C}G/M_{m_1}^{n+1}$ and $\mathbb{C}G/M_{m_2}^{n+1}$ are isomorphic.*

Proof. Observe that μ is in M_{m_1} if and only if $\Psi_{\check{m}_1 m_2}(\mu)$ is in M_{m_2}. Indeed, μ is in M_{m_1} if and only if μ annihilates m_1, that is,

$$0 = \mu(m_1) = \mu(m_2 \cdot \check{m}_2 m_1) = \Psi_{\check{m}_1 m_2}(\mu)(m_2),$$

which means, if $\Psi_{\check{m}_1 m_2}(\mu)$ annihilates m_2. It follows that

$$\Psi_{\check{m}_1 m_2}(M_{m_1}) = M_{m_2},$$

which implies

$$\Psi_{\check{m}_1 m_2}\big((M_{m_1})^{n+1}\big) = (M_{m_2})^{n+1}.$$

Now we define the mapping $F : \mathbb{C}G/M_{m_1}^{n+1} \to \mathbb{C}G/M_{m_2}^{n+1}$ by

$$F(\mu + M_{m_1}^{n+1}) = \Psi_{\check{m}_1 m_2}(\mu) + M_{m_2}^{n+1}$$

for each μ in $\mathbb{C}G$. If $\mu - \mu'$ is in $M_{m_1}^{n+1}$, then

$$\Psi_{\check{m}_1 m_2}(\mu) - \Psi_{\check{m}_1 m_2}(\mu') = \Psi_{\check{m}_1 m_2}(\mu - \mu')$$

is in $M_{m_2}^{n+1}$, as we proved above. Hence F is well-defined and clearly, it is a surjective ring homomorphism. On the other hand, if for some μ in $\mathbb{C}G$ $F(\mu + M_{m_1}^{n+1}) = M_{m_2}^{n+1}$, that is, $\Psi_{\check{m}_1 m_2}(\mu)$ is in $M_{m_2}^{n+1}$, then μ is in $M_{m_1}^{n+1}$, hence $\operatorname{Ker} F = M_{m_1}^{n+1}$, and our statement is proved. □

12.4 Generalized exponential monomials

In the case of finite Abelian groups we have seen that $\mathcal{C}(G) = L^2(G)$ always has an orthonormal basis consisting of characters, which are the common eigenfunctions of all translation operators. If G is infinite, then obviously $\mathcal{C}(G)$ is not a Hilbert space. However, it is a locally convex topological vector space, and all convolution operators, including the translation operators, are continuous linear operators. Moreover, as we have seen it in Theorem 12.7, the common eigenfunctions of all convolution operators are exactly the exponentials, in other words, the generalized characters. This property underlines the importance of exponentials. Nevertheless, it turns out that even if we replace characters by generalized characters we are still not able to perform harmonic analysis in the space of all continuous functions. Although, we have seen in Section 7.7. that in the case of a continuous function defined on a compact Abelian group those characters, which belong to $\tau(f)$ are actually take part in the reconstruction process of f, and they span a dense subspace in the variety of f. Unfortunately this is not the case on non-compact groups. For instance, as it is well-known from the theory of homogeneous linear difference equations with constant coefficients that the solution space, which is obviously a variety, is finite dimensional, however it is not spanned by either the character, or the exponential solutions. In general, the exponential solutions may have a positive multiplicity, which brings new basic functions into the picture, which should be included in the description of the variety, that is, of the solution space. At this point we arrive at the situation concerning spectral analysis and synthesis on vector modules in Section 11.4, where the fundamental role were played by matrix elements. We recall the definition in our present circumstances. Let G be a locally compact Abelian group. The continuous function $f : G \to \mathbb{C}$ is a matrix element, if it generates a finite dimensional variety. Clearly, each exponential is a matrix element. In the subsequent paragraphs we shall give different characterizations of matrix elements in terms of the annihilator of their varieties. In particular, we shall identify matrix elements with the so-called exponential polynomials. Although a part of the forthcoming results works also on arbitrary locally compact Abelian groups, in some cases we cannot avoid the use of Theorem 11.13, which holds, in general, in the discrete case only. Hence we shall almost always consider the case only, where G is a discrete Abelian group. In this case $\mathcal{M}_c(G)$ is the set of all finitely supported complex functions on G, and it can be identified with the so-called *group algebra* of G, which, in ring theory, is usually denoted by $\mathbb{C}G$. We shall use this notation. We note that, as a consequence of Theorem 11.13, in this case every maximal ideal of the group algebra is weak*-closed.

Before going into the study of exponential polynomials we introduce another important function class. If G is an Abelian group, then the function $f : G \to \mathbb{C}$ is called a *generalized exponential monomial*, if its annihilator includes a positive power of an exponential maximal ideal. In other words, there exists an exponential m and a natural number n such that we have

$$M_m^{n+1} \subseteq \tau(f)^{\perp}. \tag{12.7}$$

We show that if f is nonzero, then such an exponential is uniquely determined and it belongs to $\tau(f)$.

Lemma 12.4.1. *Let G be an Abelian group, and let V be a nonzero variety on G. Then there exists at most one maximal ideal M in $\mathbb{C}G$ such that the inclusion $M^{n+1} \subseteq V^{\perp}$ holds for some natural number n.*

Proof. As V is nonzero, hence $V^{\perp}$ is a proper ideal, hence it is included in some maximal ideal M_0. It follows $M^{n+1} \subseteq M_0$. As M_0 is maximal, it is also prime, which implies $M \subseteq M_0$, hence, by maximality, $M = M_0$. □

From this lemma it follows that for $f \neq 0$ the exponential m in (12.7) is uniquely determined, and, by the inclusion $\tau(f)^{\perp} \subseteq M_m$, we conclude $\tau(m) = M_m^{\perp} \subseteq \tau(f)$, hence m belongs to $\tau(f)$. In other words, m is the only exponential in the variety of f. Obviously, m is the only exponential included in all nonzero subvarieties of $\tau(f)$. Indeed, if V is a nonzero subvariety in $\tau(f)$, then $\tau(f)^{\perp} \subseteq V^{\perp}$. Consequently, every maximal ideal including $V^{\perp}$ also includes $\tau(f)^{\perp}$, hence it is equal to M_m. This means that spectral analysis holds for the variety of every generalized exponential monomial. If f is a nonzero generalized exponential monomial, and m is in $\tau(f)$, then we say that f *corresponds* to the exponential m, further the smallest natural number n satisfying (12.7) is called its *degree*. Although we do not define the degree of the zero function, but we consider it of degree at most n for every natural number n. The set of all generalized exponential monomials of degree at most n corresponding to the exponential m is $(M_m^{n+1})^{\perp}$, and the set of all generalized exponential monomials corresponding to the exponential m is $\bigcup_{n=0}^{n+1}(M_m^{n+1})^{\perp}$.

The following characterization of generalized exponential monomials is straightforward.

Theorem 12.10. *Let G be an Abelian group, m an exponential, and n a natural number. The function $f : G \to \mathbb{C}$ is a generalized exponential monomial of degree at most n corresponding to the exponential m if and only if*

$$\Delta_{m;y_1,y_2,\dots,y_{n+1}} * f = 0 \tag{12.8}$$

holds for each $y_1, y_2, \dots, y_{n+1}$ in G.

Proof. As the modified differences $\Delta_{m;y}$ generate M_m, hence the modified differences $\Delta_{m;y_1,y_2,\dots,y_{n+1}}$ generate M_m^{n+1}, which implies that (12.7) and (12.8) are equivalent. □

Theorem 12.11. *Let G be an Abelian group. Then the nonzero function $f : G \to \mathbb{C}$ is a generalized exponential monomial if and only if $\mathbb{C}G/\tau(f)^\perp$ is a local ring with nilpotent exponential maximal ideal.*

Proof. Let $\Phi : \mathbb{C}G \to \mathbb{C}G/\tau(f)^\perp$ be the natural homomorphism. If $f : G \to \mathbb{C}$ is a nonzero generalized exponential monomial, then we have (12.7) with a unique exponential m. As M_m is the unique maximal ideal containing $\tau(f)^\perp$, it follows that $\mathbb{C}G/\tau(f)^\perp$ is a local ring with maximal ideal $\Phi(M_m)$, moreover, by (12.7), we have

$$\Phi(M_m)^{n+1} = \Phi(M_m^{n+1}) \subseteq \Phi(\tau(f)^\perp) = \tau(f)^\perp ,$$

which means that $\Phi(M_m)$ is nilpotent. On the other hand,

$$(\mathbb{C}G/\tau(f)^\perp)/\Phi(M_m) \cong \mathbb{C}G/M_m \cong \mathbb{C} ,$$

hence $\Phi(M_m)$ is the kernel of the natural homomorphism of $\mathbb{C}G/\tau(f)^\perp$ onto $\mathbb{C}$, which is a multiplicative functional. This means that $\Phi(M_m)$ is an exponential maximal ideal.

To prove the converse, let $\mathbb{C}G/\tau(f)^\perp$ be a local ring with nilpotent exponential maximal ideal $\Phi(M)$, where M is a maximal ideal in $\mathbb{C}G$ with $\tau(f)^\perp \subseteq M$ and $\Phi(M)^{n+1} = \tau(f)^\perp$ for some natural number n. It follows that $M^{n+1} \subseteq \tau(f)^\perp$. Let $F : \mathbb{C}G/\tau(f)^\perp \to \mathbb{C}$ be a multiplicative functional with $\operatorname{Ker} F = \Phi(M)$. Then $F \circ \Phi : \mathbb{C}G \to \mathbb{C}$ is a multiplicative functional, and clearly, $\operatorname{Ker} F \circ \Phi = M$. By Theorem 12.5, we infer that $M = \tau(m)^\perp$ for some exponential m. In particular, we have $M = M_m$. Then it follows $M_m^{n+1} \subseteq \tau(f)^\perp$, hence f is a generalized exponential monomial. □

Another important property of generalized exponential monomials is expressed by the following result.

Theorem 12.12. *Let G be an Abelian group. The annihilator of each generalized exponential monomial is a primary ideal.*

Proof. We show that if f corresponds to the exponential m, then $\tau(f)^\perp$ is M_m-primary. Suppose that $M_m^{n+1} \subseteq \tau(f)^\perp \subseteq M_m$ for some exponential m and natural number n. By Krull's[3] Theorem (see [Jacobson (1989)], Vol.2, Theorem 7.1, p. 392.), the radical of $\tau(f)^\perp$ is the intersection of all prime ideals including it. If $P \supseteq \tau(f)^\perp$ is a prime ideal, then $M_m^{n+1} \subseteq P$, hence $M_m \subseteq P$, and, by maximality, $M_m = P$. It follows that the radical $\operatorname{Rad} \tau(f)^\perp$ is M_m. Now let $\mu * \nu$ be in $\tau(f)^\perp$, and suppose that ν is not in $\operatorname{Rad} \tau(f)^\perp = M_m$. Then, by the maximality of M_m, it follows that the ideal $M_m + \mathbb{C}G * \nu$, which is larger than M_m, cannot be proper, that is $M_m + \mathbb{C}G * \nu = \mathbb{C}G$. Hence $\xi + \eta * \nu = \delta_0$ holds for some ξ in M_m and η in $\mathbb{C}G$. As ξ is in the radical of $\tau(f)^\perp$, we have that ξ^k is in $\tau(f)^\perp$ for some positive integer k. Consequently, we have

$$\delta_0 = (\xi + \eta * \nu)^k = \xi^k + \theta * \nu$$

with some θ in $\mathbb{C}G$. It follows $\mu = \mu * \xi^k + \theta * \mu * \nu$, and the right hand side is in $\tau(f)^\perp$. We conclude that $\tau(f)^\perp$ is primary. □

By definition, given an exponential m on G and a natural number n the set of all generalized exponential monomials of degree at most n corresponding to m is $(M_m^{n+1})^\perp$, where $M_m = \tau(m)^\perp$. More generally, given a variety V on G and a natural number n the set of all generalized exponential monomials of degree at most n corresponding to m in V is

$$(M_m^{n+1})^\perp \cap V = (M_m^{n+1} + V^\perp)^\perp .$$

12.5 Generalized polynomials

A special class of generalized exponential monomials is formed by those corresponding to the exponential identically 1. These are called *generalized polynomials.* The definition goes back to M. Fréchet[4] and S. Mazur[5]. The study of generalized polynomials is closely related to the *Fréchet's functional equation*

$$\Delta_{1;y_1,y_2,\ldots,y_{n+1}} * f = 0\,, \tag{12.9}$$

which was studied by several mathematicians under various assumptions ([Fréchet (1909); Mazur and Orlicz (1934a,b); van der Lijn (1940a,b)]).

[3]Wolfgang Krull, German mathematician (1899-1971)
[4]Maurice Fréchet, French mathematician (1878-1973)
[5]Stanislaw Mazur, Polish mathematician (1905-1981)

M. Fréchet

S. Mazur

Generalized polynomials play a fundamental role in the theory of linear functional equations. If in (12.9) we have $y_1 = y_2 = \cdots = y_{n+1}$, then we write

$$\Delta_{1;y}^{n+1} * f = 0 \,. \tag{12.10}$$

Obviously, (12.9) implies (12.10). The converse is also true under some conditions on the domain of f. The interested reader should consult with [Székelyhidi (1991)] and the references given there ([Aczél (1966); Gajda (1984, 1987); Kuczma (2009); Székelyhidi (1979, 1982b, 1988)]. The description of generalized polynomials is possible with the help of complex homomorphisms. For the modified difference $\Delta_{1;y}$ we shall use the notation Δ_y, and we call it simply *difference.*

Let G be an Abelian group and n a positive integer. The function $F : G^n \to \mathbb{C}$ is called *n-additive*, if it is a homomorphism of G^n into the additive group of complex numbers. The 1-additive functions are called simply *additive*, and the 2-additives are called *bi-additive.* For convenience we may extend this terminology to $n = 0$ by considering $G^0 = G$, and 0-additive the constant functions. A function is called *multi-additive*, if there is a natural number n for which it is n-additive.

We note that every Abelian group can be considered as a $\mathbb{Z}$-module in the natural manner. For each positive integer k let G_k denote the tensor product of k copies of this module, and let $\Phi_k : G^k \to G_k$ denote the natural embedding given by

$$\Phi_k(x_1, x_2, \ldots, x_k) = x_1 \otimes x_2 \otimes \cdots \otimes x_k \,,$$

for each $x_1, x_2, \ldots, x_k$ in G. We also define $\operatorname{Diag}_k : G \to G^k$ by

$$\operatorname{Diag}_k(x) = (x_1, x_2, \ldots, x_k) \,,$$

where $x_1 = x_2 = \cdots = x_k = x$ for each x in G. We have the following theorem.

Theorem 12.13. *Let G be an Abelian group, n a positive integer, and let $F : G^n \to \mathbb{C}$ be a function. Then F is n-additive if and only if there exists a unique $\mathbb{Z}$-module homomorphism $A : G_n \to \mathbb{C}$ such that $F = A \circ \Phi_n$. Moreover, F is symmetric if and only if A is symmetric.*

Proof. The sufficiency is obvious. For the necessity we define

$$A(x_1 \otimes x_2 \otimes \cdots \otimes x_n) = F(x_1, x_2, \ldots, x_n),$$

whenever $x_1, x_2, \ldots, x_n$ are in G. All other statements are evident. □

To describe generalized polynomials via homomorphisms we shall need some technical tools. Using the Diag functions we define the *diagonalization*, or *trace* of a function $F : G^n \to X$, where X is any set, by $D(F) = F \circ \operatorname{Diag}_n$.

For the diagonalization of symmetric multi-additive functions we have the analogue of the *Binomial Theorem.*

Theorem 12.14. *Let G be an Abelian group, n a positive integer, and let $F : G^n \to \mathbb{C}$ be a symmetric n-additive function. Then for each x, y in G we have*

$$D(F)(x+y) = \sum_{k=0}^{n} \binom{n}{k} F\big(\operatorname{Diag}_k(x), \operatorname{Diag}_{n-k}(y)\big). \tag{12.11}$$

We note that we identify G^n with $G^k \times G^{n-k}$ in the obvious manner. In the above formula if $k = 0$, resp. $k = n$, then $\operatorname{Diag}_k(x)$, resp. $\operatorname{Diag}_{n-k}(y)$ is missing, hence

$$F\big(\operatorname{Diag}_0(x), \operatorname{Diag}_n(y)\big) = F\big(\operatorname{Diag}_n(y)\big),$$

$$F\big(\operatorname{Diag}_n(x), \operatorname{Diag}_0(y)\big) = F\big(\operatorname{Diag}_n(x)\big).$$

Proof. For $n = 1$ the statement is obvious. Proving by induction, we observe that for $n \geqslant 2$ and x, y in G we have

$$D(F)(x+y) = F\big(x+y, \operatorname{Diag}_{n-1}(x+y)\big)$$

$$= F\big(x, \operatorname{Diag}_{n-1}(x+y)\big) + F\big(y, \operatorname{Diag}_{n-1}(x+y)\big).$$

Clearly, the function $(x_1, x_2, \ldots, x_n) \mapsto F\big(z, x_1, x_2, \ldots, x_n\big)$ is symmetric and $n-1$-additive for each z in G, and its diagonalization is the function $x \mapsto F\big(z, \operatorname{Diag}_{n-1}(x)\big)$, we have, by induction

$$D(F)(x+y)$$

$$= \sum_{k=0}^{n-1} \binom{n-1}{k} \big[F(x, \mathrm{Diag}_k(x), \mathrm{Diag}_{n-1-k}(y))$$

$$+ F(y, \mathrm{Diag}_k(x), \mathrm{Diag}_{n-1-k}(y))\big]$$

$$= \sum_{k=0}^{n-1} \binom{n-1}{k} \big[F(\mathrm{Diag}_{k+1}(x), \mathrm{Diag}_{n-1-k}(y))$$

$$+ F(\mathrm{Diag}_k(x), \mathrm{Diag}_{n-k}(y))\big]$$

$$= \sum_{k=1}^{n} \binom{n-1}{k-1} \big[F(\mathrm{Diag}_k(x), \mathrm{Diag}_{n-k}(y))$$

$$+ \sum_{k=1}^{n} \binom{n-1}{k} F(\mathrm{Diag}_k(x), \mathrm{Diag}_{n-k}(y))\big]$$

$$= D(F)(x) + \sum_{k=1}^{n-1} \Big[\binom{n-1}{k-1} + \binom{n-1}{k}\Big] F\big(\mathrm{Diag}_k(x), \mathrm{Diag}_{n-k}(y)\big) + D(F)(y)$$

$$= \sum_{k=0}^{n} \binom{n}{k} F\big(\mathrm{Diag}_k(x), \mathrm{Diag}_{n-k}(y)\big)\,.$$

□

This theorem easily extends to more than two terms, and we obtain the *Multinomial Theorem.*

Theorem 12.15. *Let G be an Abelian group, n, m positive integers, and let $F : G^n \to \mathbb{C}$ be a symmetric n-additive function. Then for each $x_1, x_2, \dots, x_m$ in G we have*

$$D(F)(x_1 + x_2 + \cdots + x_m) = \tag{12.12}$$

$$\sum_{k_1+k_2+\cdots+k_m=n}^{n} \frac{n!}{k_1!k_2!\dots,k_m!} F\big(\mathrm{Diag}_{k_1}(x_1), \mathrm{Diag}_{k_2}(x_2), \dots, \mathrm{Diag}_{k_m}(x_m)\big)\,.$$

The following theorem is the *Polarization Formula.*

Theorem 12.16. *Let G be an Abelian group, m, n positive integers, and let $F : G^n \to \mathbb{C}$ be a symmetric n-additive function. Then we have for each $y_1, y_2, \dots, y_m$ in G*

$$\Delta_{y_1,y_2,\dots,y_m} * D(F) = \begin{cases} 0, & \text{whenever } m > n \\ n!\; F(y_1, y_2, \dots, y_n) & \text{whenever } m = n. \end{cases} \tag{12.13}$$

Proof. Obviously, the statement for $m = n$ implies it for $m > n$. We prove the theorem by induction on n, and it is trivial for $n = 1$. Supposing that we have proved it for all values not greater than n we prove it for $n+1$. By the elementary properties of modified differences, and by the Binomial Theorem 12.14, we have for each x in G

$$\Delta_{y_1,y_2,\dots,y_{n+1}} * D(F)(x) = \Delta_{y_1,y_2,\dots,y_n}\big[\Delta_{y_{n+1}} * D(F)\big](x) =$$

$$\Delta_{y_1,y_2,\dots,y_n} * \Big[\sum_{k=0}^{n}\binom{n+1}{k}F\big(\mathrm{Diag}_k(x), \mathrm{Diag}_{n+1-k}(y_{n+1})\big)\Big] =$$

$$\sum_{k=0}^{n}\binom{n+1}{k}\Delta_{y_1,y_2,\dots,y_n} * F\big(\mathrm{Diag}_k(x), \mathrm{Diag}_{n+1-k}(y_{n+1})\big) =$$

$$\sum_{k=1}^{n}\binom{n+1}{k}\Delta_{y_1,y_2,\dots,y_n} * F\big(\mathrm{Diag}_k(x), \mathrm{Diag}_{n+1-k}(y_{n+1})\big) =$$

$$(n+1)\; n!\; F(y_1, y_2, \dots, y_n, y_{n+1}) = (n+1)!\; F(y_1, y_2, \dots, y_n, y_{n+1}),$$

as the function $x \mapsto F\big(\mathrm{Diag}_k(x), \mathrm{Diag}_{n+1-k}(y_{n+1})\big)$ is the diagonalization of the symmetric k-additive function

$$(x_1, x_2, \dots, x_k) \mapsto F\big(x_1, x_2, \dots, x_k, \mathrm{Diag}_{n+1-k}(y_{n+1})\big). \qquad \square$$

From this we have immediately the corollaries.

Corollary 12.5.1. *Let G be an Abelian group. Then every symmetric multi-additive function is uniquely determined by its diagonalization.*

Corollary 12.5.2. *Let G be an Abelian group. For each natural number n the diagonalization of every symmetric n-additive function is a generalized polynomial of degree at most n.*

The Polarization Formula (12.13) has another important consequence: it makes possible to compute the symmetric multi-additive function from its diagonalization. This idea leads to the complete characterization of generalized polynomials using symmetric multi-additive functions.

Theorem 12.17. *Let G be an Abelian group and n a natural number. The function $f : G \to \mathbb{C}$ is a generalized polynomial of degree at most n if and only if it is the sum of diagonalizations of symmetric k-additive functions with $0 \leqslant k \leqslant n$.*

Proof. The sufficiency of the condition follows from the Polarization Formula (12.13). Suppose now that $f : G \to \mathbb{C}$ is generalized polynomial of degree at most n. Obviously, the statement is true for $n = 0$. Assuming that we have proved the statement for each natural number not greater than $n \geqslant 1$ we prove it for $n + 1$. We have

$$\Delta_{y_1,y_2,\ldots,y_{n+2}} * f = 0\,,$$

whenever $y_1, y_2, \ldots, y_{n+2}$ are in G. Let for each x in G

$$g(x) = f(x) - \frac{1}{(n+1)!}\,\Delta_x^{n+1} * f(0)\,.$$

Here Δ_x^{n+1} stands for the convolution of $n+1$ copies of Δ_x. We show that the function $F : G^{n+1} \to \mathbb{C}$ defined by

$$F_{n+1}(y_1, y_2, \ldots, y_{n+1}) = \Delta_{y_1,y_2,\ldots,y_{n+1}} * f(0)$$

is an $n+1$-additive function. Indeed, it is an easy calculation to check that we have

$$\Delta_{y+z} - \Delta_y - \Delta_z = (\delta_{-y-z} - \delta_{-y} - \delta_{-z} + \delta_0) =$$

$$(\delta_{-y} - \delta_0) * (\delta_{-z} - \delta_0) = \Delta_y * \Delta_z$$

for each y, z in G. This implies

$$F_{n+1}(y_1+z_1, y_2, \ldots, y_{n+1}) - F_{n+1}(y_1, y_2, \ldots, y_{n+1}) - F_{n+1}(z_1, y_2, \ldots, y_{n+1}) =$$

$$[\Delta_{y_1+z_1} - \Delta_{y_1} - \Delta_{z_1}] * \Delta_{y_2,\ldots,y_{n+1}} * f(0) =$$

$$\Delta_{y_1} * \Delta_{z_1} * \Delta_{y_2,\ldots,y_{n+1}} * f(0) = \Delta_{y_1,z_1,y_2,\ldots,y_{n+1}} * f(0) = 0\,,$$

which proves our statement. It follows that

$$g(x) = f(x) - \frac{1}{(n+1)!}\,D(F_{n+1})(x)$$

for each x in G. By the Polarization Formula, we have for each $x, y_1, y_2, \ldots, y_{n+1}$ in G

$$\Delta_{y_1,y_2,\ldots,y_{n+1}} * g(x) = \Delta_{y_1,y_2,\ldots,y_{n+1}} * f(x) - \Delta_{y_1,y_2,\ldots,y_{n+1}} * f(0) =$$

$$\Delta_{x,y_1,y_2,\ldots,y_{n+1}} * f(0) = 0\,,$$

that is, g is a generalized polynomial of degree at most n. Our statement follows by induction. □

Another useful result in this respect is the following (see e.g. [Djoković (1969/1970); McKiernan (1967); Székelyhidi (1982b, 1991)]).

Theorem 12.18. *Let G be an Abelian group, n a natural number, and let $f : G \to \mathbb{C}$ be a function. If*

$$\Delta_y^{n+1} * f = 0 \tag{12.14}$$

holds for each y in G, then f is a generalized polynomial of degree at most n.

Let G be an Abelian group and n a natural number. The function $f : G \to \mathbb{C}$ satisfying

$$\Delta_y^n * f = n!\ f(y) \tag{12.15}$$

for each y in G is obviously a generalized polynomial, by the previous theorem. It is called a *homogeneous* generalized polynomial. Obviously, if f is nonzero, then n is uniquely defined, and we say that f is a homogeneous generalized polynomial *of degree n*. In this case, by Theorem 12.16, we have $f = D(F)$, where $F : G \to \mathbb{C}$ is a symmetric n-additive function. Conversely, the diagonalization of a nonzero symmetric n-additive function is, obviously, a homogeneous generalized polynomial of degree n. Hence the homogeneous generalized polynomials are exactly the diagonalizations of symmetric multi-additive functions. We have the following theorem.

Theorem 12.19. *Let G be an Abelian group and n a natural number. Every generalized polynomial of degree at most n can be expressed uniquely as a sum of homogeneous generalized polynomials of degree at most n.*

Proof. We have to prove the uniqueness only. Suppose that

$$D(F_n) + D(F_{n-1}) + \cdots + D(F_0) = 0\,,$$

where F_k is k-additive and symmetric for $k = 0, 1, \ldots, n$. Then we have for each y in G

$$0 = \Delta^n_{y_1,y_2,\ldots,y_{n+1}} D(F_n) = n!\ F(y_1, y_2, \ldots, y_n)\,,$$

by Theorem 12.16. The statement follows by induction. □

If the nonzero generalized polynomial f has the representation

$$f = D(F_n) + D(F_{n-1}) + \cdots + D(F_0)\,, \tag{12.16}$$

where F_k is nonzero, k-additive, and symmetric for $k = 0, 1, \ldots, n$, then this is called the *canonical representation* of f, and $D(F_k)$ is called the *homogeneous term of degree n* of f. The canonical representation of the zero function is itself. It follows that two generalized polynomials are identical if and only if their homogeneous terms are equal. This means that when proving equality of generalized polynomials we can use the elementary method of "comparing the coefficients".

The following theorem describes the connection between generalized exponential monomials and generalized polynomials.

Theorem 12.20. *Let G be an Abelian group, n a natural number, and m an exponential. The function $f : G \to \mathbb{C}$ is a generalized exponential monomial of degree at most n corresponding to the exponential m if and only if $f = p \cdot m$, where $p : G \to \mathbb{C}$ is a generalized polynomial of degree at most n.*

Proof. The identity

$$\Delta_{m;y_1,y_2,\dots,y_{n+1}} * f(x) = m(x+y_1+y_2+\cdots+y_{n+1})\Delta_{y_1,y_2,\dots,y_{n+1}} * (f \cdot \check{m})(x)\,, \tag{12.17}$$

which holds for each $x, y_1, y_2, \dots, y_{n+1}$ in G, implies our statement. □

12.6 Generalized exponential polynomials

Finite sums of generalized exponential monomials are called *generalized exponential polynomials.* Using Theorem 12.20 we have that nonzero generalized exponential polynomials have a unique representation of the form

$$f = p_1m_1 + p_2m_2 + \cdots + p_nm_n\,, \tag{12.18}$$

where $m_1, m_2, \dots, m_n$ are different exponentials, and $p_1, p_2, \dots, p_n$ are nonzero generalized polynomials. We call this the *canonical representation* of f. The functions p_im_i belong to $\tau(f)$, by Theorem 12.22, and they are called the *monomial terms* of f. Two generalized exponential polynomials are equal if and only if their monomial terms are equal.

The following lemma will be useful.

Lemma 12.6.1. *Let G be an Abelian group, n a positive integer, k_i natural numbers, m_i exponentials for $i = 1, 2, \dots, n$, and I a proper ideal in $\mathbb{C}G$. If*

$$\Pi_{i=1}^n M_{m_i}^{k_i+1} \subseteq I\,,$$

then $I \subseteq M_{m_i}$ for $i = 1, 2, \dots, n$.

Proof. Let $V = I^\perp$. We prove by induction on n. For $n = 1$ we have

$$M_m^{k+1} \subseteq I.$$

As I is proper, it is included in a maximal ideal M. Consequently, we have $M_m^{k+1} \subseteq M$. As M is prime, we infer $M = M_m$, hence $I \subseteq M_m$. Suppose that we have proved our statement for $1, 2, \ldots, n-1$ with $n \geqslant 2$ and $\Pi_{i=1}^n M_{m_i}^{k_i+1} \subseteq I$. Then we have that $\Pi_{i=2}^n M_{m_i}^{k_i+1}$ annihilates the variety $M_{m_1}^{k_1+1}V$, that is

$$\Pi_{i=2}^n M_{m_i}^{k_i+1} \subseteq (M_{m_1}^{k_1+1}V)^\perp,$$

hence, by assumption, $(M_{m_1}^{k_1+1}V)^\perp \subseteq M_{m_i}$ for $i = 2, 3, \ldots, n$. From this we have $\tau(m_i) = (M_{m_i})^\perp \subseteq M_{m_1}^{k_1+1}V \subseteq V$, hence $I = V^\perp \subseteq \tau(m_i)^\perp = M_{m_i}$ whenever $i = 2, 3, \ldots, n$. As m_1 can be replaced by any of the m_i's, our statement follows. □

By Lemma 12.6.1 we have the following corollary.

Theorem 12.21. *Let G be an Abelian group and $f : G \to \mathbb{C}$ a generalized exponential polynomial of the form*

$$f = \varphi_1 + \varphi_2 + \cdots + \varphi_n,$$

where the φ_i's are nonzero generalized exponential monomials corresponding to different exponentials m_i. Then m_i belongs to $\tau(f)$ for $i = 1, 2, \ldots, n$.

The following theorem gives additional information in this respect.

Theorem 12.22. *Let G be an Abelian group, $n \geqslant 1$ natural number, further let $f : G \to \mathbb{C}$ be a generalized exponential polynomial of the form*

$$f = \varphi_1 + \varphi_2 + \cdots + \varphi_n,$$

where the φ_i's are generalized exponential monomials corresponding to different exponentials. Then φ_i belongs to $\tau(f)$ for $i = 1, 2, \ldots, n$.

Proof. We may suppose that all φ_i's are nonzero. By assumption, there exist different exponentials m_i and natural numbers k_i for $i = 1, 2, \ldots, n$ such that

$$M_{m_i}^{k_i+1} \subseteq \tau(\varphi_i)^\perp \subseteq M_{m_i} \quad \text{and} \quad \Pi_{j=1}^n M_{m_j}^{k_j+1} \subseteq \tau(f)^\perp$$

holds for $i = 1, 2, \ldots, n$. We prove by induction on n. Let $n = 2$. As $M_{m_1}^{k_1+1}$ and $M_{m_2}^{k_2+1}$ are co-prime ideals, the same holds, obviously, for $\tau(\varphi_1)^\perp$ and $\tau(\varphi_2)^\perp$, that is $\tau(\varphi_1)^\perp + \tau(\varphi_2)^\perp = \mathbb{C}G$. It follows that

$$\delta_0 = \mu_1 + \mu_2$$

for some μ_1 in $\tau(\varphi_1)^\perp$ and μ_2 in $\tau(\varphi_2)^\perp$. We have

$$\varphi_1 + \varphi_2 = (\mu_1 + \mu_2) * (\varphi_1 + \varphi_2) = \mu_1 * \varphi_2 + \mu_2 * \varphi_1 ,$$

hence $\varphi_1 - \mu_2 * \varphi_1 = \varphi_2 - \mu_1 * \varphi_2$. As $\tau(\varphi_1)^\perp$ and $\tau(\varphi_2)^\perp$ are co-prime, we have

$$\tau(\varphi_1) \cap \tau(\varphi_2) = \{0\} ,$$

and we infer $\varphi_1 = \mu_2 * \varphi_1$ and $\varphi_2 = \mu_1 * \varphi_2$. It follows $\mu_1 * f = \mu_1 * (\varphi_1 + \varphi_2) = \mu_1 * \varphi_2 = \varphi_2$, which is in $\tau(f)$. Similarly, φ_1 is in $\tau(f)$.

Now let $f = g + \varphi_{n+1}$, where $g = \varphi_1 + \varphi_2 + \cdots + \varphi_n$. Obviously, we have $\Pi_{j=1}^n M_{m_j}^{k_j+1} \subseteq \tau(g)^\perp$, hence, by assumption, φ_j is in $\tau(g)$ for $j = 1, 2, \ldots, n$. We show that $\tau(g)^\perp$ and $\tau(\varphi_{n+1})^\perp$ are co-prime. Indeed, supposing the contrary there is a maximal ideal M in $\mathbb{C}G$ such that

$$\tau(g)^\perp \subseteq M \ \text{ and } \ \tau(\varphi_{n+1})^\perp \subseteq M ,$$

which implies $\Pi_{j=1}^n M_{m_j}^{k_j+1} \subseteq M$. As M is prime, we have $M_j \subseteq M$ for some $1 \leqslant j \leqslant n$, and, by maximality, $M_j = M$. On the other hand, the inclusion $M_{m_{n+1}}^{k_{n+1}+1} \subseteq \tau(\varphi_{n+1})^\perp \subseteq M$ implies that $M_{m_{n+1}}^{k_{n+1}+1} \subseteq M$, thus we have again $M_{n+1} = M$, a contradiction, as $M_j \neq M_{n+1}$ for $j = 1, 2, \ldots, n$.

Finally, in the same way as for $n = 2$, we have that g and φ_{n+1} belong to $\tau(f)$, hence $\tau(g) \subseteq \tau(f)$, and our proof is complete. □

Corollary 12.6.1. *On any Abelian group nonzero generalized exponential monomials corresponding to different exponentials are linearly independent.*

Proof. We apply the previous theorem with $f = 0$. □

The following theorem characterizes generalized exponential polynomials by their annihilators.

Theorem 12.23. *Let G be an Abelian group. The function $f : G \to \mathbb{C}$ is a nonzero generalized exponential polynomial if and only if there is a positive integer k, there are natural numbers $n_1, n_2, \ldots, n_k$, and there are exponentials $m_1, m_2, \ldots, m_k$ such that we have*

$$M_{m_1}^{n_1+1} \cdot M_{m_2}^{n_2+1} \cdot \ldots \cdot M_{m_k}^{n_k+1} \subseteq \tau(f)^\perp . \tag{12.19}$$

Proof. Suppose that $f : G \to \mathbb{C}$ is a generalized exponential polynomial:

$$f = \varphi_1 + \varphi_2 + \cdots + \varphi_k ,$$

where the functions $\varphi_i : G \to \mathbb{C}$ are nonzero generalized exponential monomials, hence we have $M_{m_i}^{n_i+1} \subseteq \tau(\varphi_i)^\perp$, where m_i is an exponential $(i = 1, 2, \ldots, k)$. Let μ be a measure in $M_{m_1}^{n_1+1} \cdot M_{m_2}^{n_2+1} \cdot \cdots \cdot M_{m_k}^{n_k+1}$, then, by Theorem 11.1 and Lemma 11.1.1, we have that μ is in $M_i^{n_i+1}$ for each i, which implies $\mu(\varphi_i) = 0$ for $i = 1, 2, \ldots, k$, hence $\mu(f) = 0$. It follows that μ is in $\tau(f)^\perp$.

Conversely, suppose that (12.19) holds. Then, by Theorem 11.1 and Lemma 11.1.1, we have

$$M_{m_1}^{n_1+1} \cap M_{m_2}^{n_2+1} \cap \cdots \cap M_{m_k}^{n_k+1} \subseteq \tau(f)^\perp .$$

By Theorem 11.15, it follows

$$(M_1^{n_1+1})^\perp + (M_2^{n_2+1})^\perp + \cdots + (M_k^{n_k+1})^\perp \supseteq \tau(f) .$$

By definition, $(M_{m_i}^{n_i+1})^\perp$ consists of generalized exponential monomials for $i = 1, 2, \ldots, k$, hence our theorem is proved. □

Another characterization theorem is the following.

Theorem 12.24. *Let G be an Abelian group. The function $f : G \to \mathbb{C}$ is a generalized exponential polynomial if and only if the factor ring with respect to its annihilator is a direct sum of a finite number of local rings with nilpotent exponential maximal ideal.*

Proof. Suppose that $f : G \to \mathbb{C}$ is a generalized exponential monomial. Then, by the previous theorem we have

$$\tau(f)^\perp = \bigcap_{j=1}^{k} I_j ,$$

where $I_j = M_j^{n_j+1} + \tau(f)^\perp$. Here $M_1, M_2, \ldots, M_k$ are all the maximal ideals containing $\tau(f)^\perp$, and they have the property

$$\bigcap_{i=1}^{k} M_j^{n_j+1} \subseteq \tau(f)^\perp .$$

It follows that $\mathbb{C}G/I_j$ is a local ring with maximal ideal $M_j + I_j$, and we have $(M_j + I_j)^{n_j+1} = M_j^{n_j+1} + I_j = I_j$, hence this maximal ideal is nilpotent. We note that, by Lemma 11.1.1, the ideals I_j are pairwise co-prime. Then, by Lemma 11.1.2, (see also Theorem 1.4. in [Matsumura (1980)], p. 4.), we have

$$\mathbb{C}G/\tau(f)^\perp \cong \bigoplus_{j=1}^{k} \mathbb{C}G/I_j .$$

Conversely, suppose that we have

$$\mathbb{C}G/\tau(f)^{\perp} \cong \bigoplus_{j=1}^{k} R_j\,,$$

where R_j is a local ring with nilpotent maximal ideal. Again, by Lemma 11.1.2, it follows that we have

$$\mathbb{C}G/\tau(f)^{\perp} \cong \bigoplus_{j=1}^{k} (\mathbb{C}G/\tau(f)^{\perp})/I'_j \cong \bigoplus_{j=1}^{k} \mathbb{C}G/I_j\,,$$

where the ideals I_j are pairwise co-prime, containing $\tau(f)^{\perp}$ and their intersection is $\tau(f)^{\perp}$. As R_j is a local ring with nilpotent maximal ideal, the same holds for $\mathbb{C}G/I_j$. This means that I_j is contained in exactly one maximal ideal M_j for which $M_j^{n_j+1} \subseteq I_j$ holds with some natural number n_j ($j = 1, 2, \ldots, k$). Finally, we have, by Theorem 11.1 and Lemma 11.1.1,

$$M_1^{n_1+1} \cdot M_2^{n_2+1} \cdot \ldots \cdot M_k^{n_k+1} = \bigcap_{j=1}^{k} M_j^{n_j+1} \subseteq \bigcap_{j=1}^{k} I_j \subseteq \tau(f)^{\perp}\,,$$

and our statement follows from Theorem 12.23. □

Finally, our last characterization theorem follows.

Theorem 12.25. *Let G be an Abelian group. The function $f : G \to \mathbb{C}$ is a generalized exponential polynomial if and only if $\mathbb{C}G/\tau(f)^{\perp}$ is a semi-local ring with exponential maximal ideals and nilpotent Jacobson radical.*

Proof. Let $f : G \to \mathbb{C}$ be a generalized exponential polynomial, then, by Theorem 12.23, there is a positive integer n, there are natural numbers $k_1, k_2, \ldots, k_n$, and there exist different exponentials $m_1, m_2, \ldots, m_n$ such that

$$\Pi_{i=1}^{n} M_{m_i}^{k_i+1} \subseteq \tau(f)^{\perp}\,.$$

Suppose that M is a maximal ideal in $\mathbb{C}G$ containing $\tau(f)^{\perp}$. Then M is prime, and we have

$$\Pi_{i=1}^{n} M_{m_i}^{k_i+1} \subseteq M\,,$$

which implies that $M = M_{m_i}$ for some i. It follows that the ring $\mathbb{C}G/\tau(f)^{\perp}$ has only finitely many maximal ideals, corresponding to some of the M_{m_i}'s, hence they are exponential maximal ideals. However, by Lemma 12.6.1, we infer that actually $\tau(f)^{\perp} \subseteq M_{m_i}$ for each $i = 1, 2, \ldots, n$. This means that the Jacobson radical of $\mathbb{C}G/\tau(f)^{\perp}$ is

$$J = \bigcap_{i=1}^{n} \Phi(M_{m_i}) = \Pi_{i=1}^{n} \Phi(M_{m_i})\,,$$

where Φ is the natural homomorphism of $\mathbb{C}G$ onto $\mathbb{C}G/\tau(f)^{\perp}$. With the notation $N = \sum_{i=1} k_i + n$ we have

$$J^N = \big(\Pi_{i=1}^n \Phi(M_{m_i})\big)^N = \Phi\big(\Pi_{i=1}^n M_{m_i}^N\big) \subseteq \Phi\big(\tau(f)^{\perp}\big) = 0\,,$$

hence J is nilpotent.

Suppose now that $f \neq 0$, and $\mathbb{C}K/\tau(f)^{\perp}$ is a semi-local ring with nilpotent Jacobson radical $J = \bigcap_{i=1}^n \Phi(M_i) = \Pi_{i=1}^n \Phi(M_i)$, where Φ is the natural homomorphism of $\mathbb{C}G$ onto $\mathbb{C}G/\tau(f)^{\perp}$. Here $M_1, M_2, \ldots, M_n$ are all the maximal ideals in $\mathbb{C}G$, which include $\tau(f)^{\perp}$. Clearly, all these maximal ideal are exponential. By nilpotency, we have $J^N = 0$ for some positive integer N, which means

$$0 = \Big(\Pi_{i=1}^n \Phi(M_i)\Big)^N = \Phi(\Pi_{i=1}^n M_i^N)\,,$$

that is $\Pi_{i=1}^n M_i^N \subseteq \tau(f)^{\perp}$. Our proof is complete. □

12.7 Exponential monomials and polynomials

Although generalized exponential polynomials present an important class of functions, the basic building bricks of spectral analysis and synthesis are matrix elements, in particular, the indecomposable matrix elements. Now we introduce a subclass of generalized exponential monomials, and identify them with the indecomposable matrix elements. Let G be an Abelian group. The function $f : G \to \mathbb{C}$ is called an *exponential monomial*, if it is a generalized exponential monomial and $\tau(f)$ is finite dimensional. Exponential monomials corresponding to the exponential identically 1 are called *polynomials*. Finite sums of exponential monomials are called *exponential polynomials*. The following theorem is obvious.

Theorem 12.26. *Let G be an Abelian group. The function $f : G \to \mathbb{C}$ is an exponential polynomial if and only if it is a generalized exponential polynomial and $\tau(f)$ is finite dimensional.*

Proof. The necessity of the condition is obvious. Conversely, suppose that $f : G \to \mathbb{C}$ is a generalized exponential polynomial and $\tau(f)$ is finite dimensional. By definition, we have that f is a finite sum of generalized exponential monomials corresponding to different exponentials. Then, by Theorem 12.22, we have that these generalized exponential monomials belong to $\tau(f)$, hence they generate finite dimensional varieties, too. It follows that they are exponential monomials, and our theorem is proved. □

By Theorem 12.20, we have that the nonzero function $f : G \to \mathbb{C}$ is an exponential monomial if and only if $f = p \cdot m$, where m is an exponential and p is a polynomial, further f is an exponential polynomial if and only if $f = \sum_{i=1}^n p_i \cdot m_i$, where m_i is an exponential and p_i is a polynomial for $i = 1, 2, \dots, n$.

Theorem 12.27. *Let G be an Abelian group. The function $f : G \to \mathbb{C}$ is an exponential monomial if and only if it is an indecomposable matrix element.*

Proof. If f is a nonzero exponential monomial, then it is a generalized exponential monomial, hence, by Theorem 12.11, $\mathbb{C}G/\tau(f)^\perp$ is a local ring. Suppose that $\tau(f)$ is decomposable, then $\tau(f) = V_1 + V_2$, where V_1, V_2 are proper subvarieties of $\tau(f)$. We have $\tau(f)^\perp = V_1^\perp \cap V_2^\perp$. It follows that

$$\mathbb{C}G/\tau(f)^\perp \cong \mathbb{C}G/\Phi(V_1^\perp) \oplus \mathbb{C}G/\Phi(V_2^\perp),$$

where $\Phi : \mathbb{C}G \to \mathbb{C}G/\tau(f)^\perp$ is the natural homomorphism. The maximal ideals of the rings $\mathbb{C}G/\Phi(V_1^\perp)$ and $\mathbb{C}G/\Phi(V_2^\perp)$ can be written in the form $\Phi(M_1)$ and $\Phi(M_2)$, where M_1, M_2 are different maximal ideals containing $\tau(f)^\perp$, which is impossible, as $\mathbb{C}G\tau(f)^\perp$ is a local ring.

Let $f_1, f_2, \dots, f_n$ be a basis of the linear space $\tau(f)$, then we have

$$f_i(x + y) = \sum_{j=1}^n \lambda_{i,j}(y) f_i(x) \tag{12.20}$$

for each x, y in G with some functions $\lambda_{i,j} : G \to \mathbb{C}$, satisfying

$$\lambda_{i,j}(x + y) = \sum_{k=1}^n \lambda_{i,k}(x)\lambda_{k,j}(y) \tag{12.21}$$

for each x, y in G and for $i, j = 1, 2, \dots, n$. This is obtained exactly in the same way as (11.3) in Section 11.4. Following the argument we applied there we have that $\lambda_{i,j}(x) = \langle \Lambda(x)e_i, e_j\rangle$ holds, whenever x is in G and $i, j = 1, 2, \dots, n$, where $\Lambda : G \to GL_n(\mathbb{C})$ is a representation of G on the Hilbert–space $\mathbb{C}^n$. In particular, the functions $\lambda_{i,j}$ are matrix elements. Putting $x = 0$ in (12.20) we get immediately that f_i is a matrix element for $i = 1, 2, \dots, n$, hence f, as a linear combination of matrix elements, is a matrix element, too.

Conversely, as matrix elements span finite dimensional varieties, $\tau(f)$ is finite dimensional. Then every ascending chain of ideals in $\mathbb{C}G/\tau(f)^\perp$ induces an ascending chain of ideals containing $\tau(f)^\perp$ in $\mathbb{C}G$, which induces a descending chain of subvarieties in $\tau(f)$, by the annihilator correspondence, which terminates, as subvarieties are subspaces and $\tau(f)$ is finite dimensional. We infer that $\mathbb{C}G/\tau(f)^\perp$ is an Artin ring. As $\tau(f)$ is indecomposable, it follows that $\mathbb{C}G/\tau(f)^\perp$ is a local ring. Indeed, assuming the contrary, there are finitely many different maximal ideals $M_1, M_2, \ldots, M_k$ in $\mathbb{C}G$, which include $\tau(f)^\perp$. As $\tau(f)$ is finite dimensional, hence so are $M_1^\perp, M_2^\perp, \ldots, M_k^\perp$. Then $M_i^\perp$, as a finite dimensional vector space, includes a common eigenfunction of all translation operators, that is, there is an exponential m_i in $M_i^\perp$, which implies $M_i = \tau(m_i)^\perp$ for $i = 1, 2, \ldots, k$. Obviously, these exponentials are different. As every ideal in an Artin ring is nilpotent, we have $M_i^{n_i+1} \subseteq \tau(f)^\perp$ with some natural numbers n_i, $i = 1, 2, \ldots, k$. Then we have

$$\mathbb{C}G/\tau(f)^\perp \cong \bigoplus_{i=1}^{n} \mathbb{C}G/\Phi(M_i^{n_i+1}),$$

where $\Phi : \mathbb{C}G \to \mathbb{C}G/\tau(f)^\perp$ is the natural homomorphism. As $\tau(f)$ is indecomposable, we have $n = 1$, hence $M_1^{n_1+1} \subseteq \tau(f)^\perp$, and M_1 is an exponential maximal ideal, which means that f is a generalized exponential monomial. As $\tau(f)$ is finite dimensional, it is actually an exponential monomial, and our theorem is proved. □

In the above proof we have also verified the following result.

Corollary 12.7.1. *Let G be an Abelian group, and let V be a finite dimensional variety on G. Then $\mathbb{C}G/V^\perp$ is an Artin ring, in which every maximal ideal is exponential and nilpotent.*

Lemma 12.7.1. *Let G be an Abelian group, $f : G \to \mathbb{C}$ a function, m an exponential, and k a natural number. Then for each φ in $M_m^k\tau(f)$ the $\mathbb{C}G$-module generated by $\varphi + M_m^{k+1}\tau(f)$ in $M_m^k\tau(f)/M_m^{k+1}\tau(f)$ is one dimensional.*

Proof. Let $F : \mathbb{C}G \to \mathbb{C}$ be the multiplicative functional with the property $\operatorname{Ker}\Phi = \tau(m)^\perp$. For each y in G we have

$$\delta_{-y} * \big(\varphi + M_m^{k+1}\tau(f)\big) = \delta_{-y} * \varphi + M_m^{k+1}\tau(f) = (\delta_{-y} - m(y)\delta_0)\varphi$$
$$+ m(y)\varphi + M_m^{k+1}\tau(f) = F(\delta_{-y})\varphi + M_m^{k+1}\tau(f),$$

as $\delta_{-y} - m(y)\delta_0$ is in M_m, hence $(\delta_{-y} - m(y)\delta_0)\varphi$ is in $M_m^{k+1}\tau(f)$. As each μ in $\mathbb{C}G$ is a linear combination of measures δ_{-y}, by linearity, we have

$$\mu * (\varphi + M_m^{k+1}\tau(f) = F(\mu) \cdot \varphi + M_m^{k+1}\tau(f) =$$
$$F(\mu)\big(\varphi + M_m^{k+1}\tau(f)\big).$$

which proves our statement. □

Theorem 12.28. *Let G be an Abelian group. The function $f : G \to \mathbb{C}$ is an exponential monomial if and only if $\mathbb{C}G/\tau(f)^{\perp}$ is a local Noether ring with nilpotent exponential maximal ideal.*

Proof. The proof of the necessity is included in the proof of Theorem 12.27, where we proved that actually $\mathbb{C}G/\tau(f)^{\perp}$ is an Artin ring.

To prove the sufficiency suppose that $\mathbb{C}G/\tau(f)^{\perp}$ is a local Noether ring with nilpotent exponential maximal ideal. This implies that $\tau(f)^{\perp}$ is included in exactly one maximal ideal M such that $\big(\mathbb{C}G/\tau(f)^{\perp}\big)/\Phi(M) \cong \mathbb{C}$. This implies $\mathbb{C}G/M \cong \mathbb{C}$, hence M is exponential, and we infer $M = M_m$ for some exponential function m. On the other hand, it follows immediately $\Phi(M_m^{n+1}) = \Phi(M_m)^{n+1} = 0$ in $\mathbb{C}G/\tau(f)^{\perp}$ for some natural number n, that is, $M_m^{n+1} \subseteq \tau(f)^{\perp}$. Consequently, f is a generalized exponential monomial. We have to show only that $\tau(f)$ is finite dimensional. We may suppose that the number $n \geqslant 1$ in the previous inclusion is the smallest natural number with that property, which implies that $M_m^n\tau(f) \neq \{0\}$. Let $\varphi \neq 0$ be in $M_m^n\tau(f)$, then we have for each x, y in G

$$0 = (\delta_{-y} - m(y)\delta_0) * \varphi(x) = \varphi(x+y) - m(y)\varphi(x)\,.$$

Putting $x = 0$ we have $\varphi = \varphi(0) \cdot m$, which means that $M_m\tau(f)$ is one dimensional. We consider the chain of $\mathbb{C}G$-modules

$$\tau(f), \tau(f)/M_m\tau(f), \ldots, M_m^n\tau(f)/M_m^{n+1}\tau(f) = M_m^n\tau(f), \{0\}\,.$$

Suppose that $\tau(f)$ is infinite dimensional. Then there exists a natural number k with $0 \leqslant k \leqslant n-1$ such that $M_m^k\tau(f)/M_m^{k+1}\tau(f)$ is infinite dimensional and $M_m^{k+1}\tau(f)$ is finite dimensional. It follows that $M_m^k/M_m^{k+1}\tau(f)$ is infinite dimensional. Then there exists a sequence $\varphi_1, \varphi_2, \ldots, \varphi_l, \ldots$ in $M_m^k\tau(f)$ such that $\varphi_{l+1} + M_m^{k+1}\tau(f)$ is not included in the linear span of the elements $\varphi_j + M_m^{k+1}\tau(f)$ for $j = 1, 2, \ldots, l$ and $l = 1, 2, \ldots$. However, by Lemma 12.7.1, the linear span of the elements $\varphi_j + M_m^{k+1}\tau(f)$ for $j = 1, 2, \ldots, l$ coincides with the submodule generated by these elements in $M_m^k\tau(f)/M_m^{k+1}\tau(f)$. Consequently, φ_{l+1} is not included in the subvariety generated by the functions φ_j with $1 \leqslant j \leqslant l$ in $\tau(f)$, which means that these subvarieties form a strictly ascending chain, and their annihilators generate a strictly descending chain of ideals in $\mathbb{C}G/\tau(f)^{\perp}$. This contradicts the Noetherian property. □

Theorem 12.29. *Let G be an Abelian group. The function $f : G \to \mathbb{C}$ is an exponential monomial if and only if $\mathbb{C}G/\tau(f)^{\perp}$ is a local Artin ring with exponential maximal ideal.*

Proof. This follows from the previous results. □

Theorem 12.30. *Let G be an Abelian group. Then $f : G \to \mathbb{C}$ is an exponential polynomial if and only if $\mathbb{C}G/\tau(f)^\perp$ is an Artin ring with exponential maximal ideals. Equivalently, f is an exponential polynomial if and only if $\mathbb{C}G/\tau(f)^\perp$ is a semi-local Noether ring with exponential maximal ideals and nilpotent Jacobson radical.*

Proof. If f is an exponential monomial, then $\tau(f)$ is finite dimensional. Using the argument applied in the proof of Theorem 12.27 we deduce that $\mathbb{C}G/\tau(f)^\perp$ is an Artin ring. By the structure theorem of commutative Artin rings (see [Jacobson (1989)], Theorem 7.15, p. 426.) we have

$$\mathbb{C}G/\tau(f)^\perp \cong \bigoplus_{k=1}^{n} \mathbb{C}G/I_k\,,$$

where the $\mathbb{C}G/I_k$'s are local Artin rings. We have seen above that in this case the maximal ideals of these local rings are exponential.

Suppose now that $\mathbb{C}G/\tau(f)^\perp$ is a semi-local Noether ring with exponential maximal ideals and nilpotent Jacobson radical. Then, by Theorem 12.25, f is a generalized exponential polynomial. By Theorem 12.24, $\mathbb{C}G/\tau(f)^\perp$ is a direct sum of local rings, each having a nilpotent exponential maximal ideal. As $\mathbb{C}G/\tau(f)^\perp$ is a Noether ring, all these direct terms are Noether rings, too. In other words, we have

$$\mathbb{C}G/\tau(f)^\perp \cong \bigoplus_{k=1}^{n} \mathbb{C}G/I_k\,,$$

where $I_1, I_2, \dots, I_n$ are ideals in $\mathbb{C}G$ containing $\tau(f)^\perp$, and

$$\bigcap_{k=1}^{n} I_k = \tau(f)^\perp\,,$$

further

$$\tau(f) = \sum_{k=1}^{n} I_k^\perp\,. \tag{12.22}$$

For each φ in $I_k^\perp$ we have $I_k \subseteq \tau(\varphi)^\perp$, hence $\mathbb{C}G/\tau(\varphi)^\perp$, as a homomorphic image of $\mathbb{C}G/I_k$, is a local Noether ring, having a nilpotent exponential maximal ideal, too. By Theorem 12.28, φ is an exponential monomial, hence (12.22) implies that $\tau(f)$ consists of exponential polynomials. The theorem is proved. □

12.8 Description of exponential polynomials

In this section we give a complete description of exponential polynomials on Abelian groups using additive and multiplicative homomorphisms. We recall that homomorphisms of the Abelian group G into the additive group of complex numbers are called *additive* functions. It turns out that exponential polynomials are exactly the elements of the function algebra generated by all additive functions and exponentials. In particular, polynomials are the functions in the algebra generated by additive functions and constants. For further references concerning results in this section and related matters the reader should refer to [Stone (1960); Anselone and Korevaar (1964); Engert (1970); McKiernan (1977a,b); Laird (1979); Székelyhidi (1982a); Laird and McCann (1984); Laczkovich (2000/01, 2004)].

We shall use the following lemmas (see [McKiernan (1977b)]). Here $\mathcal{L}(\mathbb{C}^n)$ denotes the space of linear operators on $\mathbb{C}^n$, that is, the space of all $n \times n$ matrices with complex entries.

Lemma 12.8.1. *Let G be an Abelian group and n a positive integer. Let $a : G \to \mathcal{L}(\mathbb{C}^n)$ be a mapping satisfying*

$$a(x+y) = a(x) + a(x)a(y) + a(y) \tag{12.23}$$

for each x, y in G, further we assume that $a(x)$ is strictly upper triangular, that is, $a_{i,j}(x) = 0$ for $i \leqslant j$ $(i, j = 1, 2, \ldots, n)$ and x in G. Then the function $A : G \to \mathcal{L}(\mathbb{C}^n)$ defined by

$$A(x) = \sum_{k=1}^{\infty} \frac{(-1)^{n+1}}{n} a(x)^n \tag{12.24}$$

satisfies

$$A(x+y) = A(x) + A(y), \quad A(x)A(y) = A(y)A(x) \tag{12.25}$$

for each x, y in G.

We note that the sum on the right hand side in (12.24) is obviously finite, as, by assumption, $a(x)$ is strictly upper triangular, which implies $a(x)^{n+1} = 0$ for each x in G.

Proof. We note that for each x, y in G the matrices $a(x)$ and $a(y)$ commute, by (12.23). It follows that the second equation in (12.25) holds. Clearly, we have $A(x) = \ln\bigl(I + a(x)\bigr)$ for each x in G, where I is the $n \times n$ identity matrix. It follows for each x, y in G

$$A(x) + A(y) = \ln\bigl(I + a(x)\bigr) + \ln\bigl(I + a(y)\bigr) = \ln\bigl[\bigl(I + a(x)\bigr)\bigl(I + a(y)\bigr)\bigr] =$$
$$\ln\bigl(I + a(x) + a(y) + a(x)a(y)\bigr) = \ln\bigl(I + a(x+y)\bigr) = A(x+y)\,. \qquad \square$$

Lemma 12.8.2. *Let G be an Abelian group and n a positive integer. Suppose that $F : G \to \mathcal{L}(\mathbb{C}^n)$ satisfies*

$$F(x + y) = F(x)F(y)\,, \tag{12.26}$$

whenever x, y is in G, further $F(x)$ is regular for each x. Then there exists a regular matrix S in $\mathcal{L}(\mathbb{C}^n)$, and there exist positive integers $k_1, k_2, \ldots, k_l$ with $k_1 + k_2 + \cdots + k_l = n$, exponentials $m_1, m_2, \ldots, m_l$ and mappings $M_j, A_j : G \to \mathcal{L}(\mathbb{C}^{k_j})$ such that

1. *A_j is strictly upper triangular and satisfies* (12.25) *for each x, y in G,*
2. *$M_j(x) = m_j(x) \exp A_j(x)$ is upper triangular for each x in G,*
3. *$F(x) = S^{-1}\mathrm{diag}\big(M_1(x), M_2(x), \ldots, M_l(x)\big)S$ for each x in G.*

Conversely, every function $F : G \to \mathcal{L}(\mathbb{C}^n)$ having the listed properties satisfies equation (12.26).

We note that $\mathrm{diag}\big(M_1(x), M_2(x), \ldots, M_l(x)\big)$ denotes a block matrix with diagonal blocks $M_1(x), M_2(x), \ldots, M_l(x)$.

Proof. By (12.26), the matrices $M(x)$ with x in G commute. It follows (see e.g. [Newman (1967); Jacobson (1953)]) that there exists a regular matrix S such that $F(x)$ has the form given in (3) with upper triangular matrices $M_1(x), M_2(x), \ldots, M_l(x)$, and the diagonal elements of $M_j(x)$ are all equal to $m_j(x)$, where $m_j : G \to \mathbb{C}$ is nonzero for each x, because of the regularity of $M(x)$. Clearly, we have

$$M_j(x + y) = M_j(x)M_j(y) \tag{12.27}$$

for each x, y in G, hence the upper triangular property implies that each m_j is an exponential. It follows $M_j(x) = m_j(x)\big(I + a_j(x)\big)$ for each x in G, where $a_j : G \to \mathcal{L}(C^{k_j})$ is strictly upper triangular. Using equation (12.27) one easily derives that a_j satisfies (12.23). We apply the previous lemma, and infer that $A_j : G \to \mathcal{L}(C^{k_j})$ is strictly upper triangular satisfying (12.25). It follows $A_j(x) = \ln\big(I + a_j(x)\big)$ for each x in G, hence we have

$$M_j(x) = m_j(x) \exp A_j(x)$$

for each x in G, which was to be proved.

The converse statement follows by easy calculation. □

Using these auxiliary results, in the following theorem we completely describe exponential polynomials on Abelian groups with the help of additive and multiplicative homomorphisms.

Theorem 12.31. *Let G be an Abelian group, and let $f : G \to \mathbb{C}$ be an exponential polynomial. Then there exists natural numbers n, k, and for each $i = 1, 2, \dots, n; j = 1, 2, \dots, k$ there exists a polynomial $P_i : \mathbb{C}^k \to \mathbb{C}$, an exponential m_i, and a homomorphism a_j of G into the additive group of complex numbers such that*

$$f(x) = \sum_{i=1}^{n} P_i\big(a_1(x), a_2(x), \dots, a_k(x)\big) m_i(x) \tag{12.28}$$

holds for each x in G. Conversely, every function of the given form is an exponential polynomial.

Proof. By Theorem 12.20 it is obviously enough to prove the corresponding statement for polynomials, in which case $n = 1$ and $m = 1$ in (12.28).

First we prove the sufficiency of the given condition. Hence we suppose that

$$f(x) = P\big(a_1(x), a_2(x), \dots, a_k(x)\big)$$

holds for each x in G, where $P : \mathbb{C}^k \to \mathbb{C}$ is an arbitrary polynomial, and $a_1, a_2, \dots, a_k : G \to \mathbb{C}$ are additive functions. Then, by Taylor's[6] Formula, we have for each x in G

$$f(x + y) = P\big(a_1(x) + a_1(y), a_2(x) + a_2(y), \dots, a_k(x) + a_k(y)\big) =$$

$$\sum \frac{1}{\alpha_1! \dots \alpha_k!} \partial_1^{\alpha_1} \dots \partial_k^{\alpha_k} P\big(a_1(x), \dots, a_k(x)\big) a_1(y)^{\alpha_1}, \dots, a_k(y)^{\alpha_k},$$

where the summation extends to all multi-indices $\alpha = (\alpha_1, \alpha_2, \dots, \alpha_k)$ in $\mathbb{N}^k$ with

$$|\alpha| = \alpha_1 + \alpha_2 + \dots + \alpha_k \leqslant \deg P .$$

This shows that $\tau(f)$ is generated by the finite set of functions

$$\{\partial_1^{\alpha_1} \dots \partial_k^{\alpha_k} P\big(a_1(x), \dots, a_k(x)\big) : \ |\alpha| \leqslant \deg P\},$$

hence it is finite dimensional, which implies that f is a polynomial.

Suppose now that $f : G \to \mathbb{C}$ is a polynomial. It follows that $\tau(f)$ is finite dimensional, and the only exponential function in $\tau(f)$ is $m = 1$. Let $f_1, f_2, \dots, f_n$ be a basis of $\tau(f)$, then there exists functions $\lambda_{i,j} : G \to \mathbb{C}$ such that

$$f_i(x + y) = \sum_{j=1}^{n} \lambda_{i,j}(y) f_j(x) \tag{12.29}$$

[6] Brook Taylor, English mathematician (1685-1731)

holds for each x, y in G. Using the linear independence of the functions $f_1, f_2, \ldots, f_n$ we infer that

$$\lambda_{i,j}(x+y) = \sum_{k=1}^{n} \lambda_{i,k}(x)\lambda_{k,j}(y) \tag{12.30}$$

holds for each x, y in G. Again, by the linear independence of the functions $f_1, f_2, \ldots, f_n$, there exist elements $x_1, x_2, \ldots, x_n$ in G such that the matrix $\big(f_i(x_j)\big)_{i,j=1}^{n}$ is regular. Let y be in G, and let i be arbitrary with $1 \leqslant i \leqslant n$. Substituting $x = x_k$ in (12.30) we get a system of linear equations for the unknowns $\lambda_{i,j}$:

$$f_i(x_k+y) = \sum_{j=1}^{n} \lambda_{i,j}(y) f_j(x_k) \tag{12.31}$$

for $k = 1, 2, \ldots, n$. The fundamental matrix of this linear system of equations is regular, hence, by Cramer's Rule[7], $\lambda_{i,j}(y)$ is a linear combination of some translates of f_i, that is, each $\lambda_{i,j}$ is in $\tau(f)$.

For each x in G let $F(x)$ denote the $n \times n$ matrix $(\lambda_{i,j})_{i,j=1}^{n}$, then $F(x)$ is regular, and, by (12.30), we have

$$F(x+y) = F(x)F(y) \tag{12.32}$$

for each x, y in G. Using Lemma 12.8.2 we have

$$F(x) = S^{-1}\mathrm{diag}\big(m_1(x)\exp A_1(x), m_2(x)\exp A_2(x), \ldots, m_l(x)\exp A_l(x)\big)S$$

for each x in G, where $A_j : G \to \mathcal{L}(\mathbb{C}^{k_j})$ is strictly upper triangular satisfying equation (12.25), further $m_j : G \to \mathbb{C}$ is an exponential for $j = 1, 2, \ldots, l$.

G. Cramer

B. Taylor

[7]Gabriel Cramer, Swiss mathematician (1704-1752)

Obviously, the entries of A_j are homomorphisms of G into $\mathbb{C}$. It follows that the entries of F, that is, the functions $\lambda_{i,j}$ are of the form (12.28). Substituting $x = 0$ in (12.30) we have that the functions f_i, hence all functions in $\tau(f)$ have the form (12.28), too, and our theorem is proved.□

12.9 An example

In this section we give an example for an Abelian group and a generalized exponential monomial on it, which is not an exponential monomial. In what follows for each cardinality κ the symbol $\mathbb{Z}^{(\kappa)}$ denotes the *non-complete direct product*, or *infinite direct sum* of κ copies of $\mathbb{Z}$, that is, the set of all $\mathbb{Z}$-valued finitely supported functions on a set of cardinality κ. Let $G = \mathbb{Z}^{(\omega)} = \bigoplus_{n\in\mathbb{N}}\mathbb{Z}$. We shall use the *projections* $p_n : \mathbb{Z}^{(\omega)} \to \mathbb{Z}$ for $n = 0, 1, \ldots$, where $p_n(x) = x(n)$ for each x in $\mathbb{Z}^{(\omega)}$ and n in $\mathbb{N}$. We define the function

$$B(x, y) = \sum_{n\in\mathbb{N}} p_n(x)p_n(y)$$

for each x, y in G. Clearly, the sum is finite for each x, y, hence the function $B : G \times G \to \mathbb{C}$ is well-defined, and symmetric. It is obvious that we have

$$B(x_1 + x_2, y) = B(x_1, y) + B(x_2, y)\,, \tag{12.33}$$

whenever x_1, x_2, y are in G. Now we let $f(x) = B(x, x)$ for each x in G. Using equation (12.33) it is straightforward to verify that

$$f(x + y) = f(x) + 2B(x, y) + f(y)$$

for each x, y in G. It follows that

$$(\delta_{-y} - \delta_0) * f(x) = f(x + y) - f(x) = 2B(x, y) + f(y)\,, \tag{12.34}$$

$$(\delta_{-y} - \delta_0) * (\delta_{-z} - \delta_0) * f(x) = 2B(y, z)\,, \tag{12.35}$$

and

$$(\delta_{-y} - \delta_0) * (\delta_{-z} - \delta_0) * (\delta_{-w} - \delta_0) * f(x) = 0 \tag{12.36}$$

for each x, y, z, w in G.

From (12.35) it follows that all constant functions belong to $\tau(f)$. Then $\tau(f)^{\perp} \subseteq M_1$, and here the subscript refers to the exponential identically 1. As M_1 is generated by all measures $\delta_{-y} - \delta_0$ for y in G, it follows that M_1^3 is generated by all measures of the form

$$(\delta_{-y} - \delta_0) * (\delta_{-z} - \delta_0) * (\delta_{-w} - \delta_0)$$

with y, z, w in G. By (12.36) it follows, that M_1^3 annihilates $\tau(f)$, hence we have

$$M_1^3 \subseteq \tau(f)^\perp \subseteq M_1\,,$$

that is, f is a generalized exponential monomial.

As all constant functions belong to $\tau(f)$, it follows, by (12.34), that the functions $x \mapsto B(x, y)$ belong to $\tau(f)$ for each y in G. In particular, if $e_k(n) = 1$, or 0 depending on $n = k$, or $n \neq k$, then we have

$$B(x, e_k) = \sum_{n \in \mathbb{N}} p_n(x) p_n(e_k) = p_n(x)\,,$$

which shows that p_n belongs to $\tau(f)$ for each n in $\mathbb{N}$. As the projections are obviously linearly independent, hence $\tau(f)$ is infinite dimensional. We infer that f is a generalized exponential monomial, which is not an exponential monomial.

This example can be used to show that there exists a generalized exponential monomial, which is not an exponential monomial on each Abelian group, which has a subgroup isomorphic to $\mathbb{Z}^{(\omega)}$.

Theorem 12.32. *If an Abelian group has a subgroup isomorphic to* $\mathbb{Z}^{(\omega)}$*, then there is a generalized exponential monomial on it, which is not an exponential monomial.*

Proof. We may obviously suppose that $\mathbb{Z}^{(\omega)}$ is itself a subgroup of the Abelian group G. For each n in $\mathbb{N}$ let p_n denote the projection of the direct sum $\mathbb{Z}^{(\omega)} = \oplus_{n \in \mathbb{N}} \mathbb{Z}$ onto the n-th copy of $\mathbb{Z}$. Clearly, each p_n is a homomorphism of $\mathbb{Z}^{(\omega)}$ into the additive group of complex numbers. As it is divisible, by Theorem 5.3, each p_n has an extension to a homomorphism of G into $\mathbb{C}$, hence the same construction we have seen in the previous example provides a generalized exponential monomial on G, which is not an exponential monomial. $\square$

Chapter 13

THE TORSION FREE RANK

13.1 Basics from group theory

It turns out that the torsion properties of the underlying group play an important role in spectral analysis and spectral synthesis on Abelian groups. The first general result in this respect is in [Székelyhidi (2001)] about spectral analysis on torsion groups, which was followed by the basic theorem in [Laczkovich and Székelyhidi (2005)], where the torsion free rank of the group is used to characterize the presence of spectral analysis on the group. It turns out that the torsion free rank has a close connection to the existence of generalized exponential monomials, which are not exponential monomials, and exactly these functions may violate spectral synthesis. This is illustrated also by the results of [Bereczky and Székelyhidi (2005)], where spectral synthesis is proved on torsion groups.

In this chapter we present the basic concepts concerning the torsion free rank of Abelian groups, its properties and connections with polynomials and generalized polynomials.

Let G be an Abelian group. The *rank* of G is the smallest cardinality of a generating set of G. An element x in G is called of *finite order*, or a *torsion element*, if it generates a finite subgroup. This is exactly the case if there is a positive integer n with $n \cdot x = 0$. The smallest n with this property is called the *order* of x. An element, which is not of finite order is called an element of *infinite order*. Hence x is an element of infinite order if and only if the subgroup generated by x is isomorphic to the additive group of integers. We call G a *torsion group*, or *periodic group*, if every element of G is of finite order. We call G *torsion free*, if every nonzero element of G is of infinite order.

The following theorem shows that the subset of all torsion elements in an Abelian group form a subgroup, which is called the *torsion subgroup* of the group, and the corresponding factor group is torsion free (see Theorem (A.4) in [Hewitt and Ross (1979)], p. 440.).

Theorem 13.1. *Let G be an Abelian group. All torsion elements of G form a subgroup T, and G/T is torsion free.*

Proof. Let T be the set of all torsion elements of G, then for x, y in T we have positive integers m, n such that $m \cdot x = n \cdot y = 0$, hence $mn \cdot (x-y) = 0$, which implies that $x - y$ is in T, and T is a subgroup. Suppose that x is not in T, and n is a positive integer such that $n \cdot (x+T) = T$. This implies $n \cdot x$ is in T, hence there is a positive integer m with $m \cdot (n \cdot x) = 0$. We have $(m \cdot n) \cdot x = 0$, hence x is in T, a contradiction. This shows that G/T is torsion free. □

We call the *torsion free rank* of the Abelian group G the *cardinality* of a maximal *linearly independent system* of elements. Independence is meant over the integers, that is, the set $x_1, x_2, \ldots, x_n$ of elements is called *linearly independent*, if an equation

$$k_1 x_1 + k_2 x_2 + \cdots + k_n x_n = 0$$

with some integers $k_1, k_2, \ldots, k_n$ implies $k_1 = k_2 = \cdots = k_n = 0$. Obviously, the elements of any linearly independent set are of infinite order. Instead of *linearly independent* we shall simply say *independent*. For more about the torsion free rank see e.g. [Fuchs (1958, 1970); Hewitt and Ross (1979)]. We have the following characterization.

Theorem 13.2. *The torsion free rank of an Abelian group G is κ if and only if G has a subgroup H isomorphic to $\mathbb{Z}^{(\kappa)}$ such that G/H is a torsion group.*

Proof. Suppose that the torsion free rank of G is $\kappa \neq 0$, then there is an independent subset $(x_i)_{i \in I}$ of elements in G such that for each x in G the set $\{x_i, x, i \in I\}$ is dependent. Here I is a set of cardinality κ. Let H denote the subgroup generated by the x_i's, and let x be an arbitrary element in G. If x is of finite order, then there is a positive integer m such that $mx = 0$, hence mx is in H. If x is of infinite order, then, by the maximality of the system $(x_i)_{i \in I}$, there exist elements $x_{i_1}, x_{i_2}, \ldots, x_{i_k}$ and nonzero integers $m_1, m_2, \ldots, m_k, m$ such that

$$mx = \sum_{j=1}^{k} m_j x_{i_j} .$$

It follows that mx is in H. We infer that for each x in G there exists a positive integer m such that mx is in H. Hence, in the factor group G/H we have

$$m \cdot (x + H) = mx + H = H\,,$$

that is, the factor group is a torsion group.

Now let $\Phi : H \to \mathbb{Z}^I$ be defined as follows: for each h in H we have

$$h = \sum_{j=1}^{k} m_j x_{i_j} \tag{13.1}$$

with some integers m_j, $j = 1, 2, \ldots, k$, and this representation of h is unique, by the independence of the set $(x_i)_{i \in I}$. Then we let

$$\Phi(h)(i_j) = m_j \ \text{ for } j = 1, 2, \ldots, k\,,$$

and $\Phi(h)(i) = 0$, whenever i is different from each i_j with $j = 1, 2, \ldots, k$. The function Φ is well-defined, and it is clearly an isomorphism of H onto the group $\mathbb{Z}^{(\kappa)}$, the infinite direct sum of κ copies of $\mathbb{Z}$.

Conversely, if G has a subgroup H isomorphic to $\mathbb{Z}^{(\kappa)}$, and G/H is a torsion group, then H has an independent set $(x_i)_{i \in I}$ of generators, where I is a set of cardinality κ. For each element x in G there is a nonzero integer m such that $m \cdot (x + H) = H$, that is, mx is in H, which means that the set $\{x, x_i, i \in I\}$ is dependent, hence $(x_i)_{i \in I}$ is a maximal independent set in G, and its cardinality is κ, hence the torsion free rank of G is κ. □

It turns out that the torsion free rank of G is infinite if and only if it has a subgroup isomorphic to the group $\mathbb{Z}^{(\omega)}$. Indeed, if G has a subgroup isomorphic to $\mathbb{Z}^{(\omega)}$, then it contains the characteristic function of the set $\{n\}$ for each n in $\mathbb{N}$, and these functions are clearly independent, hence the torsion free rank of G is at least ω. On the other hand, if the torsion free rank is at least ω, then for each n in $\mathbb{N}$ the group G has a subgroup H_n isomorphic to $\mathbb{Z}^n$, and the subgroup generated by $\bigcup_{n \in \mathbb{N}} H_n$ contains a subgroup isomorphic to the group $\mathbb{Z}^{(\omega)}$.

For an arbitrary Abelian group G we use the notation $\mathrm{Hom}\,(G, \mathbb{C})$ for the linear space of all complex homomorphisms of G into the additive group of complex numbers. We recall that $\mathrm{Hom}\,(G, \mathbb{C})$ is the set of all additive functions on G. The following result enlightens the connection of the dimension of $\mathrm{Hom}\,(G, \mathbb{C})$ to the torsion free rank of G (see also in [Székelyhidi (2005)]).

Theorem 13.3. *Let G be an Abelian group. The torsion free rank of G is finite if and only if* Hom$(G,\mathbb{C})$ *is a finite dimensional vector space, and in this case its dimension is equal to the torsion free rank of G.*

Proof. Suppose that the torsion free rank of G is n, where n is a positive integer, and let H denote a subgroup of G isomorphic to $\mathbb{Z}^n$, such that G/H is a torsion group. Let $\Phi : H \to \mathbb{Z}^n$ be an isomorphism. If $p_i : \mathbb{Z}^n \to \mathbb{Z}$ denotes the projection of $\mathbb{Z}^n$ onto the i-th copy of $\mathbb{Z}$ for $i = 1, 2, \ldots, n$, then we let

$$a_i(x) = p_i\big(\Phi(x)\big)$$

for $i = 1, 2, \ldots, n$ and for each x in H. Then $a_i : H \to \mathbb{C}$ is a complex homomorphism, and clearly, the a_i's are linearly independent. By Theorem 5.3, each a_i has an extension A_i to a homomorphism of G into $\mathbb{C}$, and these extensions are obviously linearly independent, too. Let $A : G \to \mathbb{C}$ be an arbitrary complex homomorphism of G. The function $z \mapsto A\big(\Phi^{-1}(z)\big)$ on $\mathbb{Z}^n$ is obviously a complex homomorphism. As it is easy to see, Hom$(\mathbb{Z}^n, \mathbb{C})$ is an n-dimensional vector space, and the projections obviously form a basis in it, hence we have

$$A\big(\Phi^{-1}(z)\big) = \sum_{i=1}^{n} \lambda_i p_i(z)$$

for each z in $\mathbb{Z}^n$ with some complex numbers λ_i, $i = 1, 2, \ldots, n$. For each x in H we put $z = \Phi(x)$ in this equation to obtain

$$A(x) = \sum_{i=1}^{n} \lambda_i a_i(x)\,.$$

Let x be arbitrary in G. As G/H is a torsion group, hence there exists a positive integer k such that $k \cdot (x + H) = H$, whence kx is in H. From the previous equation we have

$$k \cdot A(x) = A(kx) = \sum_{i=1}^{n} \lambda_i a_i(kx) = \sum_{i=1}^{n} \lambda_i A_i(kx) = k \cdot \sum_{i=1}^{n} \lambda_i A_i(x)\,,$$

and division by k gives that A is a linear combination of the functions $A_1, A_2, \ldots, A_n$, hence Hom$(G,\mathbb{C})$ is an n-dimensional vector space.

Conversely, let $a_1, a_2, \ldots, a_n$ be a basis of Hom$(G,\mathbb{C})$. Then there are elements x_j in G for $j = 1, 2, \ldots, n$ such that the matrix $\big(a_i(x_j)\big)_{i,j=1}^{n}$ is regular. It is obvious that these elements are independent, hence the torsion free rank of G is at least n. For each natural number k with $1 \leqslant k \leqslant n$ there are complex numbers $\lambda_{1,k}, \lambda_{2,k}, \ldots, \lambda_{n,k}$ such that

$$\sum_{j=1}^{n} \lambda_{j,k} a_j(x_i) = 0 \tag{13.2}$$

for $1 \leqslant i \leqslant n$ and $i \neq k$, further

$$\sum_{j=1}^{n} \lambda_{j,k} a_j(x_k) = 1 \,. \tag{13.3}$$

We define for each x in G

$$b_k(x) = \sum_{j=1}^{n} \lambda_{j,k} a_j(x) \,.$$

Then b_k is in $\mathrm{Hom}\,(G, \mathbb{C})$ for $k = 1, 2, \ldots, n$, further $b_k(x_i) = 1$ or 0 depending on $k = i$ or $k \neq i$. Hence the functions b_k are linearly independent for $k = 1, 2, \ldots, n$. Suppose now that there is an x_{n+1} in G such that $x_1, x_2, \ldots, x_n, x_{n+1}$ are independent, and let H be the subgroup of G generated by these elements. We define $b : H \to \mathbb{C}$ in the following manner. For each h in H there are unique integers $m_1, m_2, \ldots, m_{n+1}$ such that

$$h = \sum_{j=1}^{n+1} m_j x_j \,.$$

The existence follows from the fact that the x_j's generate H, and the uniqueness is the consequence of their independence. Then we can define $b(h) = m_{n+1}$. It is easy to see that b is in $\mathrm{Hom}\,(H, \mathbb{C})$. By Theorem 5.3, b has an extension to a homomorphism B of G into $\mathbb{C}$, then B is in $\mathrm{Hom}\,(G, \mathbb{C})$, and there are complex numbers $\lambda_1, \lambda_2, \ldots, \lambda_n$ such that

$$B(h) = \sum_{j=1}^{n} \lambda_j b_j(h) \tag{13.4}$$

holds for each h in H. Then putting $h = x_i$ for $i = 1, 2 \ldots, n$ we have that $0 = B(x_i) = b(x_i) = \lambda_i$ for $i = 1, 2, \ldots, n$, which implies, by (13.4), that B vanishes on H. However, $B(x_{n+1}) = b(x_{n+1}) = 1$, a contradiction. This proves that the x_j's form a maximal independent set in G, hence the torsion free rank of G is n. The theorem is proved. □

We have the following simple corollaries.

Corollary 13.1.1. *Every subgroup and every homomorphic image of an Abelian group of finite torsion free rank has finite torsion free rank, too. In particular, every factor group of an Abelian group of finite torsion free rank has finite torsion free rank, too. The direct sum of two Abelian groups of finite torsion free rank has finite torsion free rank, too.*

Proof. The first statement follows immediately from the definition. Suppose that G has finite torsion free rank, and $\Phi : G \to H$ is a surjective homomorphism. If $a_1, a_2, \ldots, a_n : H \to \mathbb{C}$ are linearly independent complex homomorphisms, then obviously, $a_1 \circ \Phi, a_2 \circ \Phi, \ldots, a_n \circ \Phi : G \to \mathbb{C}$ are linearly independent complex homomorphisms, too, which implies the second statement, by the previous theorem. The third statement is obvious, as every factor group is a homomorphic image. Finally, we note that

$$\mathrm{Hom}\,(G \oplus H, \mathbb{C}) = \mathrm{Hom}\,(G, \mathbb{C}) \oplus \mathrm{Hom}\,(H, \mathbb{C}),$$

which implies the last statement. □

Corollary 13.1.2. *Every finitely generated Abelian group has finite torsion free rank.*

Proof. Every finitely generated Abelian group is the homomorphic image of $\mathbb{Z}^k$ for some natural number k. □

13.2 The torsion free rank and polynomials

The following theorem shows that the existence of generalized polynomials, which are not polynomials is closely related to the finiteness of the torsion free rank.

Theorem 13.4. *The torsion free rank of an Abelian group is finite if and only if every generalized polynomial on the group is a polynomial.*

Proof. Suppose that the torsion free rank of the Abelian group G is finite. If it is zero, then the statement is obvious, because in that case any complex generalized polynomial on G is a constant. Hence we suppose that the torsion free rank of G is the positive integer k, and then, by the Theorem 13.3, any additive function is a linear combination of some fixed linearly independent additive functions $a_1, a_2, \ldots, a_k$. First we describe the general form of the multi-additive functions on G. If $B : G \times G \to \mathbb{C}$ is bi-additive then $x \mapsto B(x, y)$ is additive for each y in G, hence there are functions $\lambda_1, \lambda_2, \ldots, \lambda_k : G \to \mathbb{C}$ such that

$$B(x, y) = \lambda_1(y)a_1(x) + \lambda_2(y)a_2(x) + \cdots + \lambda_k(y)a_k(x)$$

holds for any x, y in G. The linear independence of the functions $a_1, a_2, \ldots, a_k$ implies that there are elements $g_1, g_2, \ldots, g_k$ in G such that the $k \times k$ matrix $\big(a_i(g_j)\big)$ is regular. Substituting $x = g_j$ for $j = 1, 2, \ldots, k$

into the above equation we get a linear system of equations from which it is clear that the functions $\lambda_1, \lambda_2, \ldots, \lambda_k$ are linear combinations of the functions $y \mapsto B(g_1, y)$ for $j = 1, 2, \ldots, k$, hence they are additive. This means that these functions are also linear combinations of the functions $a_1, a_2, \ldots, a_k$. Therefore the general form of the bi-additive functions on G is the following:

$$B(x, y) = \sum_{i=1}^{k} \sum_{j=1}^{k} b_{ij} a_i(x) a_j(y)\,,$$

where b_{ij} are complex numbers for $i, j = 1, 2, \ldots, k$. Repeating this argument we get by induction that for each positive integer n the general form of the n-additive functions on G is the following:

$$A(x_1, x_2, \ldots, x_n) = \sum_{i_1=1}^{k} \sum_{i_2=1}^{k} \cdots \sum_{i_n=1}^{k} m_{i_1 i_2 \ldots i_n}\, a_{i_1}(x_1) a_{i_2}(x_2) \ldots a_{i_n}(x_n)\,,$$

where $m_{i_1 i_2 \ldots i_n}$ are complex numbers for $i_1, i_2, \ldots, i_n = 1, 2, \ldots, k$. Finally, by Theorem 12.17, we infer that on G every generalized polynomial is a polynomial.

Conversely, suppose that the torsion free rank of G is infinite. Let $X = \{x_i : i \in \mathbb{N}\}$ be an independent sequence of elements in G. Then for every i in $\mathbb{N}$ there is a homomorphism $p_i : G \to \mathbb{C}$ such that $p_i(x_j) = 1$ for $i = j$ and $p_i(x_j) = 0$ for $i \neq j$. Indeed, first we find such a homomorphism on the subgroup H generated by X, then we extend it to G. It follows that the functions p_i are linearly independent for different values i in $\mathbb{N}$. Finally, we let

$$B(x, y) = \sum_{i \in \mathbb{N}} p_i(x) p_i(y)$$

for each x, y in G. The sum is finite for each x, y in G. Indeed, by the maximality of X, for each g in G there is a positive integer n such that ng belongs to H. Then

$$ng = m_1 x_{j_1} + m_2 x_{j_2} + \cdots + m_k x_{j_k}$$

with some integers $m_1, m_2, \ldots, m_k$ and elements $x_{j_1}, x_{j_2}, \ldots, x_{j_k}$ in X. It is clear that $p_i(g) = 0$ for every $i \neq j_1, j_2, \ldots, j_k$.

Obviously, B is a symmetric and bi-additive function. On the other hand, if x_j is any element of X then we have

$$B(x, x_j) = \sum_{i \in \mathbb{N}} p_i(x) p_i(x_j) = p_j(x)\,,$$

hence the functions $x \mapsto B(x, x_j)$ are linearly independent for different values i in $\mathbb{N}$. On the other hand, we have seen in Section 12.9 that all these functions belong to the translation invariant linear space generated by the function $x \mapsto B(x, x)$, hence this linear space is of infinite dimension. It follows immediately that the generalized polynomial $x \mapsto B(x, x)$ is not a polynomial. □

In fact, we have proved the following theorem.

Theorem 13.5. *The torsion free rank of an Abelian group is finite if and only if every bi-additive function is a bi-linear function of additive functions.*

Here "bi-linear" means that it has the form $(x, y) \mapsto \sum_{k=1}^{n} a_k(x)b_k(y)$, where a_k, b_k are additive functions for $k = 1, 2, \ldots, n$.

Using Theorem 12.20 we have the following corollaries.

Corollary 13.2.1. *The torsion free rank of an Abelian group is finite if and only if every generalized exponential monomial on the group is an exponential monomial.*

Corollary 13.2.2. *The torsion free rank of an Abelian group is finite if and only if every generalized exponential polynomial on the group is an exponential polynomial.*

13.3 The polynomial ring

We introduce the polynomial ring on Abelian groups (see [Székelyhidi (2005, 2012a,b)]).

Theorem 13.6. *All generalized polynomials on an Abelian group form a ring.*

Proof. Clearly, the sum of two generalized polynomials is a generalized polynomial, too. For the rest of the proof, by Theorem 12.17, it is enough to show that the product of the diagonalizations of an m-additive and an n-additive function is the diagonalization of an $m + n$-additive function. Let $f : G^m \to \mathbb{C}$ and $g : G^n \to \mathbb{C}$ be m-additive, respectively n-additive functions, then the function $h : G^{m+n} \to \mathbb{C}$ defined by,

$$h(x_1, x_2, \ldots, x_m, y_1, y_2, \ldots, y_n) = f(x_1, x_2, \ldots, x_m)g(y_1, y_2, \ldots, y_n)$$

for $x_1, x_2, \ldots, x_m, y_1, y_2, \ldots, y_n$ in G is clearly $m+n$-additive. It is obvious, that the diagonalization of h is the product of the diagonalizations of f and g. This implies our statement. □

The ring of all generalized polynomials on the Abelian group G is called the *polynomial ring* of G and is denoted by $\mathcal{P}(G)$.

Theorem 13.7. *Let G be an Abelian group and let T denote its torsion subgroup. Then the polynomial rings of G and G/T are isomorphic.*

Proof. Let $\Phi : G \to G/T$ denote the natural homomorphism and we define

$$F(p) = p \circ \Phi$$

for each p in $\mathcal{P}(G/T)$. We show that $F(p)$ is in $\mathcal{P}(G)$. As p is in $\mathcal{P}(G/T)$, there exists a natural number n such that

$$\Delta_{v_1, v_2, \ldots, v_{n+1}} * p(u) = 0$$

holds for each $u, v_1, v_2, \ldots, v_{n+1}$ in G/T. Let $x, y_1, y_2, \ldots, y_{n+1}$ be arbitrary in G, then we have

$$\Delta_{y_1, y_2, \ldots, y_{n+1}} * F(p)(x) = \Delta_{y_1, y_2, \ldots, y_{n+1}} * p(\Phi(x)) =$$

$$\Delta_{\Phi(y_1), \Phi(y_2), \ldots, \Phi(y_{n+1})} * p(\Phi(x)) = 0\,.$$

□

The following theorem relates to Theorem 13.2.

Theorem 13.8. *Let G be an Abelian group, and let H be a free subgroup such that G/H is a torsion group. Then the polynomial rings of G and H are isomorphic.*

Proof. It is enough to show that every generalized polynomial on H has a unique extension to a generalized polynomial on G. The statement, that every generalized polynomial on H has an extension to a generalized polynomial on G has been proved in [Székelyhidi (2000)], hence we just give a hint for the proof of this statement here. In fact, in the light of Theorem 12.17, it is enough to show that every k-additive function on H^k has an extension to a k-additive function on G^k for each positive integer on k. For this we can use Theorem 5.3 about the extension of additive functions.

To complete the proof we have to show that the extension of generalized polynomials from H to G is unique. Again, using Theorem 12.17, it is enough to show this about the extension of multi-additive functions. Suppose that k is a positive integer and $A, B : G^k \to \mathbb{C}$ are k-additive functions satisfying

$$A(h_1, h_2, \ldots, h_k) = B(h_1, h_2, \ldots, h_k)$$

whenever h_j is in H for $j = 1, 2, \ldots, k$. Let $x_1, x_2, \ldots, x_k$ be in G. As G/H is a torsion group, there exist positive integers $n_1, n_2, \ldots, n_k$ such that $n_j x_j$ is in H for $j = 1, 2, \ldots, k$. Hence we have

$$n_1 \cdot n_2 \cdot \cdots \cdot n_k A(x_1, x_2, \ldots, x_k) = A(n_1x_1, n_2x_2, \ldots, n_kx_k) =$$

$$B(n_1x_1, n_2x_2, \ldots, n_kx_k) = n_1 \cdot n_2 \cdot \cdots \cdot n_k B(x_1, x_2, \ldots, x_k),$$

which implies $A(x_1, x_2, \ldots, x_k) = B(x_1, x_2, \ldots, x_k)$, which proves our theorem. □

The following theorem characterizes Abelian groups of finite torsion free rank in terms of the polynomial ring (see [Székelyhidi (2012a)]).

Theorem 13.9. *Let G be an Abelian group. Then the polynomial ring of G is Noetherian if and only if the torsion free rank of G is finite.*

Proof. Using Theorem 13.3 it is enough to show that the polynomial ring of G is Noetherian if and only if the dimension of $\mathrm{Hom}\,(G, \mathbb{C})$ is finite. Suppose first that $\mathrm{Hom}\,(G, \mathbb{C})$ is infinite dimensional. We show that if the additive function a is not a linear combination of the additive functions $a_1, a_2, \ldots, a_n$, then a is not included in the ideal I_n generated by the polynomial functions $a_1, a_2, \ldots, a_n$. Supposing the contrary there exist polynomial functions $p_1, p_2, \ldots, p_n$ in $\mathcal{P}(G)$ such that

$$a(x) = \sum_{i=1}^{n} p_i(x) a_i(x) \tag{13.5}$$

holds for each x in G. If N is an upper bound for the degrees of the polynomial functions p_i, $i = 1, 2, \ldots, n$, then we have the representation

$$p_i(x) = \sum_{j=0}^{N} A_j^{(i)}(x)$$

for $i = 1, 2, \ldots, n$ and for each x in G, where $A_j^{(i)} : G \to \mathbb{C}$ is zero, or it is a homogeneous generalized polynomial of degree j for $j = 0, 1, \ldots, N$ and $i = 1, 2, \ldots, n$. Substituting into (13.5) we get

$$a(x) = \sum_{j=0}^{N} \sum_{i=1}^{n} A_j^{(i)}(x) a_i(x) \tag{13.6}$$

for all x in G. It is obvious that the function $x \mapsto A_j^{(i)}(x)a_i(x)$ is either zero, or a monomial function of degree $j+1$ for $j = 0, 1, \ldots, N$. As the left hand side is a homogeneous generalized polynomial of degree 1, it follows, by the uniqueness, that

$$\sum_{i=1}^{n} A_j^{(i)}(x)a_i(x) = 0$$

for $j = 1, 2, \ldots, N$ and

$$\sum_{i=1}^{N} A_0^{(i)} a_i(x) = a(x)$$

holds for each x in G, which contradicts our assumption.

As the dimension of $\operatorname{Hom}(G, \mathbb{C})$ is infinite, there exists a sequence $(a_n)_{n\in\mathbb{N}}$ of linearly independent additive functions on G. If I_n denotes the ideal in $\mathcal{P}(G)$ generated by the polynomials $a_0, a_1, \ldots, a_n$ for $n = 0, 1, \ldots,$ then, by our previous assertion, the chain $I_0 \subset I_1 \subset \cdots$ is strictly ascending, hence $\mathcal{P}(G)$ is not Noetherian.

Now suppose that $\dim \operatorname{Hom}(G, \mathbb{C}) = n < +\infty$. We show that the polynomial ring $\mathcal{P}(G)$ of G is isomorphic to the polynomial ring $\mathbb{C}[z_1, z_2, \ldots, z_n]$, hence it is Noetherian.

Let $\{a_1, a_2, \ldots, a_n\}$ be a basis of the space $\operatorname{Hom}(G, \mathbb{C})$. Our first step is to prove that each k-additive function $F : G^k \to \mathbb{C}$ has a representation in the form

$$F(x_1, x_2, \ldots, x_k) = \sum_{i_1=1}^{n} \sum_{i_2=1}^{n} \cdots \sum_{i_k=1}^{n} \lambda_{i_1,i_2,\ldots,i_k} a_{i_1}(x_1)a_{i_2}(x_2)\ldots a_{i_k}(x_k), \tag{13.7}$$

where $x_1, x_2, \ldots, x_k$ are in G, and $\lambda_{i_1,i_2,\ldots,i_k}$ are complex numbers for $i_1, i_2, \ldots, i_k = 1, 2, \ldots, n$. This is obviously true for $k = 1$. We suppose that it is true for $k \geqslant 1$, and let $F : G^{k+1} \to \mathbb{C}$ be a $k+1$-additive function. For each x in G the function

$$(x_1, x_2, \ldots, x_k) \mapsto F(x_1, x_2, \ldots, x_k, x)$$

is k-additive, hence it has a representation in the form given in (13.7), that is, there are complex numbers $\lambda_{i_1,i_2,\ldots,i_k}(x)$ – depending on x – such that

$$F(x_1, x_2, \ldots, x_k, x) = \sum_{i_1=1}^{n} \sum_{i_2=1}^{n} \cdots \sum_{i_k=1}^{n} \lambda_{i_1,i_2,\ldots,i_k}(x) a_{i_1}(x_1)a_{i_2}(x_2)\ldots a_{i_k}(x_k), \tag{13.8}$$

holds for each $x_1, x_2, \ldots, x_k, x$ in G. As the additive functions $a_1, a_2, \ldots, a_n$ are linearly independent, it follows from [Székelyhidi (1999)], Theorem 3.2.7 on p. 33., that the functions $(x_1, x_2, \ldots, x_k) \mapsto a_{i_1}(x_1)a_{i_2}(x_2)\ldots a_{i_k}(x_k)$ are linearly independent for different choices of $i_1, i_2, \ldots, i_k = 1, 2, \ldots, n$. This means, that for any choice of the i_k's there exist elements $x_1^{(i_1)}$, $x_2^{(i_2)}, \ldots, x_k^{(i_k)}$ in G such that the quadratic matrix built up from the numbers $a_{i_1}(x_1^{(j_1)})a_{i_2}(x_2^{(j_2)})\ldots a_{i_k}(x_k^{(j_k)})$ is regular. Substituting these values into (13.8) we get the system of linear equations

$$F(x_1^{(j_1)}, x_2^{(j_2)}, \ldots, x_k^{(j_k)}, x) \tag{13.9}$$

$$= \sum_{i_1=1}^{n} \sum_{i_2=1}^{n} \cdots \sum_{i_k=1}^{n} \lambda_{i_1, i_2, \ldots, i_k}(x) a_{i_1}(x_1^{(j_1)}) a_{i_2}(x_2^{(j_2)}) \ldots a_{i_k}(x_k^{(j_k)})$$

for $j_1, j_2, \ldots, j_k = 1, 2, \ldots, n$. By using Cramer's Rule we can express $\lambda_{i_1, i_2, \ldots, i_k}(x)$ from this system as a linear combination of the $F(x_1^{(j_1)}, x_2^{(j_2)}, \ldots, x_k^{(j_k)}, x)$'s. As the latter expressions are additive in x, hence $\lambda_{i_1, i_2, \ldots, i_k}$ is additive for $i_1, i_2, \ldots, i_k = 1, 2, \ldots, n$ and we have the desired expression for F.

From this part it follows that each generalized polynomial on G is a polynomial, that is, it has the form

$$p(x) = P\big(a_1(x), a_2(x), \ldots, a_n(x)\big) \tag{13.10}$$

for all x in G, with some complex polynomial P in $\mathbb{C}[z_1, z_2, \ldots, z_n]$. We show that the representation of p in the form (13.10) is unique. Suppose, that there is another polynomial Q in $\mathbb{C}[z_1, z_2, \ldots, z_n]$ such that

$$p(x) = P\big(a_1(x), a_2(x), \ldots, a_n(x)\big) = Q\big(a_1(x), a_2(x), \ldots, a_n(x)\big) \tag{13.11}$$

holds for each x in G. By Taylor's Formula we have

$$P(\xi_1, \xi_2, \ldots, \xi_n) = \tag{13.12}$$

$$\sum_{|\alpha_1|+|\alpha_2|+\cdots+|\alpha_n| \leqslant N} \frac{1}{\alpha_1! \alpha_2! \ldots \alpha_n!} \partial_1^{\alpha_1} \partial_2^{\alpha_2} \ldots \partial_n^{\alpha_n} P(0, 0, \ldots, 0) \xi_1^{\alpha_1} \xi_2^{\alpha_2} \ldots \xi_n^{\alpha_n},$$

and similarly

$$Q(\xi_1, \xi_2, \ldots, \xi_n) = \tag{13.13}$$

$$\sum_{|\alpha_1|+|\alpha_2|+\cdots+|\alpha_n| \leqslant N} \frac{1}{\alpha_1! \alpha_2! \ldots \alpha_n!} \partial_1^{\alpha_1} \partial_2^{\alpha_2} \ldots \partial_n^{\alpha_n} Q(0, 0, \ldots, 0) \xi_1^{\alpha_1} \xi_2^{\alpha_2} \ldots \xi_n^{\alpha_n}$$

for all $\xi_1, \xi_2, \ldots, \xi_n$ in $\mathbb{C}$. Here N is the maximum of the degrees of P and Q. We can use the multi-index notation: for each $\alpha = (\alpha_1, \alpha_2, \ldots, \alpha_n)$ in $\mathbb{N}^n$ we write $|\alpha| = |\alpha_1| + |\alpha_2| + \cdots + |\alpha_n|$, $\alpha! = \alpha_1!\alpha_2!\ldots\alpha_n!$, further $\partial^\alpha = \partial_1^{\alpha_1}\partial_2^{\alpha_2}\ldots\partial_n^{\alpha_n}$. Then we have

$$P\big(a_1(x), a_2(x), \ldots, a_n(x)\big) = \tag{13.14}$$

$$\sum_{|\alpha|\leqslant N} \frac{1}{\alpha!}\partial^\alpha P(0,0,\ldots,0)a_1(x)^{\alpha_1}a_2(x)^{\alpha_2}\ldots a_n(x)^{\alpha_n},$$

and

$$Q\big(a_1(x), a_2(x), \ldots, a_n(x)\big) = \tag{13.15}$$

$$\sum_{|\alpha|\leqslant N} \frac{1}{\alpha!}\partial^\alpha Q(0,0,\ldots,0)a_1(x)^{\alpha_1}a_2(x)^{\alpha_2}\ldots a_n(x)^{\alpha_n}$$

for all x in G. As the additive functions $a_1, a_2, \ldots, a_n$ are linearly independent, by the theorem in [Székelyhidi (1999)] cited above, it follows that also the functions

$$x \mapsto a_1(x)^{\alpha_1}a_2(x)^{\alpha_2}\ldots a_n(x)^{\alpha_n}$$

are linearly independent for $|\alpha| \leqslant N$. Then there are elements x_β in G for each $\beta = (\beta_1, \beta_2, \ldots, \beta_n)$ in $\mathbb{N}^n$ with $|\beta| \leqslant N$ such that the quadratic matrix built up from the numbers $a_1(x_\beta)^{\alpha_1}a_2(x_\beta)^{\alpha_2}\ldots a_n(x_\beta)^{\alpha_n}$ with $|\alpha| \leqslant N, |\beta| \leqslant N$ is regular. Substituting x_β for x in (13.14), respectively (13.15) for each β with $|\beta| \leqslant N$ we obtain systems of linear equations for the unknowns $\partial^\alpha P(0,0,\ldots,0)$, respectively $\partial^\alpha Q(0,0,\ldots,0)$ with $|\alpha| \leqslant N$. By Cramer's Rule these unknowns are linear combinations of the left hand sides. On the other hand, the left hand sides are equal to $p(x_\beta)$ in the case of both systems, hence

$$\partial^\alpha P(0,0,\ldots,0) = \partial^\alpha Q(0,0,\ldots,0)$$

holds for each α with $|\alpha| \leqslant N$. This implies that $P = Q$.

We have proved that equation (13.10) sets up a well-defined mapping between the rings $\mathcal{P}(G)$ and $\mathbb{C}[z_1, z_2, \ldots, z_n]$. Clearly, this mapping is a bijective ring-isomorphism, hence our theorem is proved. □

Chapter 14

SPECTRAL ANALYSIS

14.1 Review on basics

In this chapter we study the problem of spectral analysis for varieties on discrete Abelian groups. In this introductory section we summarize the basic concepts and terminology related to this subject.

Let G be an Abelian group, and let V be a variety on G. According to the definition given in Section 11.5 we say that spectral analysis holds for V, if every nonzero subvariety of V has a nonzero finite dimensional subvariety. In particular, spectral analysis holds for the zero variety. More generally, spectral analysis holds for every finite dimensional variety. We say that spectral analysis holds on G, if spectral analysis holds for each variety on G, or equivalently, if spectral analysis holds for $\mathcal{C}(G)$. By the results in Chapter 11.6, in particular, in Section 12.7, we have the following theorem.

Theorem 14.1. *Let G be an Abelian group, and let V be a variety on G. Then the following statements are equivalent.*

1. *There is a nonzero finite dimensional subvariety in V.*
2. *There is an exponential in V.*
3. *There is a nonzero exponential monomial in V.*
4. *There is a nonzero exponential polynomial in V.*

Proof. Obviously, we have the implications (2) $\Rightarrow$ (3) $\Rightarrow$ (4). If V has a nonzero finite dimensional subvariety V_0, then V_0 is a common invariant subspace of all translation operators. These operators form a commuting family of linear operators in $\mathcal{L}(V_0)$, hence they have a common eigenfunction in V_0, which is, by Theorem 12.1, an exponential in V. It follows that (1) implies (2). Finally, if there is a nonzero exponential polynomial in V, then, by definition, it spans a nonzero finite dimensional subvariety in V, consequently (4) implies (1). □

We have the following simple corollary.

Corollary 14.1.1. *Let G be an Abelian group, and let V be a variety on G. Then the following statements are equivalent.*

1. *Spectral analysis holds for V.*
2. *Each nonzero subvariety of V contains an exponential.*
3. *Each nonzero subvariety of V contains a nonzero exponential monomial.*
4. *Each nonzero subvariety of V contains a nonzero exponential polynomial.*

We also have the following results as easy consequences.

Corollary 14.1.2. *If spectral analysis holds on an Abelian group, then it holds on every subgroup of it, too.*

Proof. Let G be an Abelian group on which spectral analysis holds, and let H be a subgroup. If V is a nonzero variety on H, and

$$W = \{f : f \text{ is in } \mathcal{C}(G) \text{ and } \tau_y f\big|_H \text{ is in } V \text{ for each } y \text{ in } G\},$$

then V is a variety on G. Let U be a subset of G containing exactly one element in each coset of H. If $f \neq 0$ is in V, further we define $g(u+x) = f(x)$ for u in U and x in H, then obviously g is in W, hence W is nonzero. By assumption, W contains an exponential m. Then $m\big|_H$ belongs to V, hence spectral analysis holds on H. □

Corollary 14.1.3. *If spectral analysis holds on an Abelian group, then it holds on every homomorphic image of it, too.*

Proof. Let G be an Abelian group on which spectral analysis holds, and let $\Phi : G \to H$ be a surjective homomorphism. If V is a nonzero variety on H, then

$$W = \{f \circ \Phi : f \text{ is in } V\}$$

is a nonzero variety on G, as it is easy to see. By spectral analysis, W contains an exponential e, and $e = m \circ \Phi$ for some function m in V. We have for each u, v in H with $u = \Phi(x)$ and $v = \Phi(y)$ the equation

$$m(u+v) = m\big(\Phi(x) + \Phi(y)\big) = m\big(\Phi(x+y)\big) =$$
$$e(x+y) = e(x)e(y) = m\big(\Phi(x)\big)m\big(\Phi(y)\big) = m(u)m(v),$$

which means that m is an exponential in V, and the proof is complete. □

The following theorem characterizes varieties possessing spectral analysis in terms of their annihilator.

Theorem 14.2. *Let G be an Abelian group, and let V be a variety on G. Spectral analysis holds for V if and only if every maximal ideal including its annihilator is exponential.*

Proof. This follows immediately from Theorem 12.6. □

This theorem has the following obvious reformulation.

Theorem 14.3. *Let G be an Abelian group, and let V be a variety on G. Spectral analysis holds for V if and only if every maximal ideal of the ring $\mathbb{C}G/V^\perp$ is exponential.*

Proof. Indeed, in the proof of Theorem 12.11 we have shown that the maximal ideal M with $V^\perp \subseteq M$ in $\mathbb{C}G$ is exponential if and only if the maximal ideal $\Phi(M)$ in $\mathbb{C}G/V^\perp$ is exponential, where $\Phi : \mathbb{C}G \to \mathbb{C}G/V^\perp$ is the natural homomorphism. □

This theorem shows that spectral analysis for a given variety V depends on the ring theoretical properties of the algebra $\mathbb{C}G/V^\perp$. In particular, spectral analysis on G depends on the ring theoretical properties of the group algebra $\mathbb{C}G$.

Corollary 14.1.4. *Spectral analysis holds on an Abelian group if and only if every maximal ideal of its group algebra is exponential.*

14.2 Spectral analysis and torsion properties

Obviously, an Abelian group has torsion free rank zero if and only if it is a torsion group. In this case spectral analysis holds on the group, as the following theorem shows (see [Székelyhidi (2001)]).

Theorem 14.4. *Spectral analysis holds on every Abelian torsion group.*

Proof. Let V be any nonzero variety on G. We show that V contains a character of G. Let Γ be a finite subset of the annihilator $V^\perp$ of V, and we let $V_\Gamma = \Gamma^\perp$. We show that for each nonempty finite subset $\Gamma \subseteq V^\perp$ the variety V_Γ contains a character. Indeed, let F_Γ denote the subgroup generated by the supports of the measures belonging to Γ. Then F_Γ is a finitely generated torsion group, hence it is a finite Abelian group. The measures belonging to Γ can be considered as measures on F_Γ, and the annihilator of Γ in $\mathcal{C}(F_\Gamma)$ will be denoted by V_{F_Γ}. This is a variety on F_Γ. It is also nonzero. Indeed, if f belongs to V, then its restriction to F_Γ belongs to V_{F_Γ}. If, in addition, we have $f(x_0) \neq 0$ and y_0 is in F_Γ, then the translate of f by $x_0 - y_0$ belongs to V, its restriction to F_Γ belongs to V_{F_Γ}, and at y_0 it takes the value $f(x_0) \neq 0$. Hence V_{F_Γ} is a nonzero variety on F_Γ. As F_Γ is finite, spectral analysis holds for V_{F_Γ}. By Corollary 14.1.1, it contains an exponential. As F_Γ is finite, every exponential on F_Γ is bounded, hence it is a character. This means that V_{F_Γ} contains a character of F_Γ. By Theorem 5.3, every character of F_Γ can be extended to a character of G, and obviously any such extension belongs to V_Γ.

Now we have proved that for each finite subset $\Gamma \subseteq \Lambda$ the variety V_Γ contains a character. Let $char(V)$ denote the set of all characters contained in V. Obviously $char(V)$ is a compact subset of $\widehat{G}$, the dual of G, because $char(V)$ is closed, and $\widehat{G}$ is compact. On the other hand, the family of nonempty compact sets $char(V_\Gamma)$, where $\Gamma \subseteq \Lambda$ is a finite subset, is centered:

$$char(V_{\Gamma_1 \cup \Gamma_2}) \subseteq char(V_{\Gamma_1}) \cap char(V_{\Gamma_2}).$$

We infer that the intersection of this family is nonempty, and obviously

$$\varnothing \neq \bigcap_{\Gamma \subseteq \Lambda \ \text{finite}} char(V_\Gamma) \subseteq char(V).$$

This means that $char(V)$ is nonempty, and the theorem is proved. □

A fundamental result in spectral analysis is due to M. Laczkovich and G. Székelyhidi, who proved in [Laczkovich and Székelyhidi (2005)] that it holds on a discrete Abelian group if and only if the torsion free rank of the group is less than the continuum (see Theorem 14.5 below). For the proof of this theorem they used the following lemma, which is a variant of [Matsumura (1986)], Theorem 5.2, p. 32. and is contained in Problem 10(b) of [Bourbaki (1972)], Chapter V. §3, p. 373.

Lemma 14.2.1. *Suppose that K is a field, k is a subfield, and X is a subset of K such that $\max(|X|, \omega) < |k|$, and $K = k[X]$; that is, K, as a ring, is generated by k and X. Then K is an algebraic extension of k.*

Theorem 14.5. *Spectral analysis holds on an Abelian group if and only if its torsion free rank is less than the continuum.*

Proof. First we prove the sufficiency. Suppose that G is an Abelian group and its torsion free rank satisfies $r_0(G) < 2^\omega$. By Theorem 14.3, it is enough to show that every maximal ideal M in $\mathbb{C}G$ is exponential, or, what is the same, that the field $K = \mathbb{C}G/M$ is isomorphic to the complex field. We prove this in the subsequent paragraphs.

Let $F : \mathbb{C}G \to \mathbb{C}G/M$ denote the natural homomorphism. Clearly, $k = \{c \cdot \delta_0 : c \in \mathbb{C}\}$ is a subfield in $\mathbb{C}G$, which is isomorphic to $\mathbb{C}$. Then $c \cdot \delta_0 \mapsto F(c \cdot \delta_0)$ is an isomorphism from k into K, as $c \neq 0$ implies that $c \cdot \delta_0$ is not in M, hence $F(c \cdot \delta_0) \neq 0$ in K. It follows that $F(k)$ is a subfield in K isomorphic to $\mathbb{C}$, and if we identify it with $\mathbb{C}$, then K is an extension of $\mathbb{C}$. Let for each x in G

$$m(x) = F(\delta_x),$$

then we have for each x, y in G

$$m(x + y) = F(\delta_{x+y}) = F(\delta_x * \delta_y) = F(\delta_x)F(\delta_y) = m(x)m(y).$$

Obviously, K is the ring generated by $\mathbb{C}$ and by the elements $m(x)$ with x in G, in other words $K = \mathbb{C}[m(G)]$.

Let T denote the torsion subgroup of G, then, by Theorem 13.1, G/T is torsion free. Let X be a subset in G containing exactly one element from each coset of T. We show that the cardinality $|X|$ of X satisfies $|X| < 2^\omega$. Supposing the contrary, by assumption, G/T contains a free Abelian group of rank $\geqslant 2^\omega$. Let Y be a set of generators of this free group with $|Y| \geqslant 2^\omega$. Let $\psi : G \to G/T$ denote the natural homomorphism, and select a point from each set $\psi^{-1}(y)$ with y in Y. Since the set Z of these points is linearly independent, it follows that the torsion free rank of G is at least $|Z| \geqslant 2^\omega$, a contradiction. Consequently, $|X| < 2^\omega$.

Every element g in G can be written in the form $g = x + t$ with x in X and t in T. For t in T there is a natural number n with $nt = 0$, hence $1 = m(0) = m(nt) = m(t)^n$, and we infer that $m(t)$ is an n-th root of unity. In particular, $m(t)$ is a complex number for each t in T, that is, $m(T)$ is in $\mathbb{C}$. It follows that $K = \mathbb{C}[m(X)]$, and here $|m(X)| \leqslant |X| < 2^\omega = |\mathbb{C}|$. By Lemma 14.2.1, we have that K is an algebraic extension of $\mathbb{C}$. As $\mathbb{C}$ is algebraically closed, we conclude that $K = \mathbb{C}$.

For the necessity we note that if the torsion free rank of G is not less than 2^ω, then G contains a free subgroup of rank 2^ω. Hence, by Corollary 14.1.2, it is enough to show that spectral analysis does not hold on the free Abelian group of rank 2^ω. Although this is just a reformulation of Problem 10(c) of [Bourbaki (1972)], Chapter V. §3, p. 373., here is a direct proof by showing that if G is the free Abelian group of rank 2^ω, then there is a maximal ideal M in $\mathbb{C}G$ such that $\mathbb{C}G/M$ is not isomorphic to the complex field (see [Laczkovich and Székelyhidi (2005)]). Indeed, let X be the set of generators of G, then $|X| = 2^\omega$. There is a surjective mapping $\phi : X \to \mathbb{C}(t)$, $\mathbb{C}(t)$ being the quotient field of the complex polynomial ring. Then there is a unique homomorphism $F : \mathbb{C}G \to \mathbb{C}(t)$ with $F(\delta_x) = \phi(x)$ for each x in X. As $\mathbb{C}(t)$ is a field, hence $\operatorname{Ker} F$ is a maximal ideal in $\mathbb{C}G$. However, $\mathbb{C}G/\operatorname{Ker} F \equiv \mathbb{C}(t)$, hence it is not isomorphic to $\mathbb{C}$. This means that not every maximal ideal in $\mathbb{C}G$ is exponential, hence, by Corollary 14.1.4, spectral analysis fails to hold on G. Our proof is complete. □

Chapter 15

SPECTRAL SYNTHESIS

15.1 Review on basics and history

In this chapter we study the problem of spectral synthesis for varieties on discrete Abelian groups, and this introductory section is devoted to the basic concepts and terminology related to the subject.

Let G be an Abelian group, and let V be a variety on G. According to the definition given in Section 11.5 we say that V is synthesizable, if the finite dimensional subvarieties of V span a dense subvariety in V. In particular, every finite dimensional variety is synthesizable. We say that spectral synthesis holds for V, if every subvariety of it is synthesizable. Clearly, spectral synthesis holds for every finite dimensional variety. We say that spectral synthesis holds on G, if spectral synthesis holds for each variety on G, or equivalently, if spectral synthesis holds for $\mathcal{C}(G)$. Obviously, spectral synthesis for a variety implies spectral analysis for it.

The first general result on non-discrete spectral synthesis was published in [Schwartz (1947)], where the following theorem was proved.

Theorem 15.1. *(L. Schwartz) Spectral synthesis holds on the reals.*

This beautiful theorem can be considered as the starting point of the research in spectral analysis and spectral synthesis. In 1954 B. Malgrange[1] proved the following result (see [Malgrange (1954)]).

Theorem 15.2. *(B. Malgrange) For any nonzero linear partial differential operator $P(D)$ in $\mathbb{R}^n$ spectral synthesis holds for the solution space of the partial differential equation $P(D)f = 0$.*

[1]Bernard Malgrange, French mathematician (1928-)

This result has been generalized by L. Ehrenpreis[2] in 1955 by proving the following theorem (see [Ehrenpreis (1955)]). We recall that an ideal is called *principal*, if it is generated by a single element. In the theorem $\mathcal{E}(\mathbb{C}^n)$ denotes the Schwartz space of infinitely differentiable complex valued functions on $\mathbb{C}^n$.

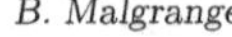

B. Malgrange

L. Ehrenpreis

Theorem 15.3. *(L. Ehrenpreis) Spectral synthesis holds for each variety in $\mathcal{E}(\mathbb{C}^n)$, whose annihilator is a principal ideal.*

The first general result on discrete Abelian groups, published in 1958 by M. Lefranc, was the following theorem (see [Lefranc (1958)]).

Theorem 15.4. *(M. Lefranc) Spectral synthesis holds on $\mathbb{Z}^n$.*

In 1965 R. J. Elliott published a result (see [Elliott (1965)]) claiming that spectral synthesis holds on every Abelian group. Unfortunately, in 1987 Z. Gajda pointed out that Elliott's proof had a gap. Finally, 17 years later it turned out that not just Elliott's proof was defective, but, in fact, his theorem was false. We shall come back to this point later.

In 1975 D. I. Gurevič presented the following result (see [Gurevič (1975)]), which was the first negative result in the non-discrete case.

Theorem 15.5. *(D. I. Gurevič) Spectral synthesis fails to hold on $\mathbb{R}^n$, if $n \geqslant 2$.*

Using the results on exponential monomials and exponential polynomials in Chapter 11.6, in particular in Section 12.7, we have the following theorem.

[2]Leon Ehrenpreis, American mathematician (1930-2010)

Theorem 15.6. *Let G be an Abelian group, and let V be a variety on G. Then the following statements are equivalent.*

1. *V is synthesizable.*
2. *The exponential monomials in V span a dense subspace.*
3. *The exponential polynomials in V span a dense subspace.*

The following two theorems can be proved using the same ideas we applied in Theorems 14.1.2 and 14.1.3.

Theorem 15.7. *If spectral synthesis holds on an Abelian group, then it holds on every subgroup of it, too.*

Theorem 15.8. *If spectral synthesis holds on an Abelian group, then it holds on every homomorphic image of it, too.*

15.2 Synthesizable varieties

The following theorems characterize synthesizable varieties in terms of their annihilator.

Theorem 15.9. *Let G be an Abelian group, and let V be a variety on G. Then V is synthesizable if and only if $V^\perp = \bigcap_{V^\perp \subseteq I} I$, where the intersection is extended to all ideals I containing $V^\perp$ such that $\mathbb{C}G/I$ is a local Artin ring with exponential maximal ideal.*

Proof. By Theorem 11.15, the given condition is equivalent to the following:

$$V = \sum_{V^\perp \subseteq I} I^\perp$$

assuming that $\mathbb{C}G/I$ is a local Artin ring with exponential maximal ideal. Clearly, by Theorem 12.29, this is equivalent to the property, that all exponential monomials in V span a dense subspace, which is exactly the synthesizability of V. □

Theorem 15.10. *Let G be an Abelian group, and let V be a variety on G. Then V is synthesizable if and only if $\mathbb{C}G/V^\perp$ is embedded into a direct product of local Artin rings with exponential maximal ideal.*

Proof. If V is synthesizable, then it is the topological sum of all subvarieties generated by exponential monomials belonging to V, by definition.

This means that we have, by Theorem 11.15,

$$V^{\perp} = \bigcap_{\varphi \in V} \tau(\varphi)^{\perp}, \tag{15.1}$$

where the intersection is extended to all exponential monomials φ in V. We define the mapping

$$F : \mathbb{C}G \to \Pi_{\varphi} \mathbb{C}G/\tau(\varphi)^{\perp}$$

by

$$F(\mu)(\varphi) = \mu + \tau(\varphi)^{\perp}$$

for each exponential monomial φ in V. Then, clearly, F is a ring homomorphism of $\mathbb{C}G$ into the direct product $\Pi_{\varphi}\mathbb{C}G/\tau(\varphi)^{\perp}$. The kernel of F consists of those μ in $\mathbb{C}G$ belonging to $\tau(\varphi)^{\perp}$ for each exponential monomial φ in V, that is, by (15.1), the kernel of F is $V^{\perp}$. It follows that $\mathbb{C}G/V^{\perp}$ is isomorphic to its image by F, which is a subring of the direct product of local Artin rings, by Theorem 12.29.

The converse is obvious. □

Now we investigate non-synthesizable varieties. The following result is clear, by Theorem 11.15.

Theorem 15.11. *Let G be an Abelian group. A variety on G is decomposable if and only if its annihilator is the intersection of two ideals, which are different from it.*

It follows that if the annihilator of an indecomposable variety is the intersection of a family of ideals, then it is equal to some member of the family.

Theorem 15.12. *Let G be an Abelian group and V a variety on G. If V is indecomposable, and spectral synthesis holds for V, then $\mathbb{C}G/V^{\perp}$ is a local Artin ring.*

Proof. If V is synthesizable, then, by Theorem 15.9, we have

$$V^{\perp} = \bigcap_{V^{\perp} \subseteq I} I, \tag{15.2}$$

where the intersection is extended to all ideals I containing $V^{\perp}$ such that $\mathbb{C}G/I$ is a local Artin ring with exponential maximal ideal. As V is indecomposable, hence, by Theorem 15.11, $V^{\perp} = I$ for some ideal I with $\mathbb{C}G/I$ is a local Artin ring. □

Theorem 15.13. *Let G be an Abelian group, and let $f : G \to \mathbb{C}G$ be a generalized exponential monomial. Then $\tau(f)$ is synthesizable if and only if f is an exponential monomial.*

Proof. The statement is obvious by the definition of exponential monomials and by the previous theorem. □

The following corollary follows immediately.

Corollary 15.2.1. *Let G be an Abelian group, and let V be a variety on G. If V contains a generalized exponential monomial, which is not an exponential monomial, then spectral synthesis fails to hold for V.*

Corollary 15.2.2. *If spectral synthesis holds on an Abelian group, then every generalized exponential polynomial on this group is an exponential polynomial.*

Using Corollary 13.2.2 we have the failure of spectral synthesis on Abelian groups with infinite torsion free rank. This was proved originally in [Székelyhidi (2004)].

Corollary 15.2.3. *Spectral analysis fails to hold on any Abelian group of infinite torsion free rank.*

The above results suggest a strong connection between spectral synthesis and the non-existence of "pathological" polynomials on an Abelian group. A reasonable question arises (see [Székelyhidi (2004)]): is the finiteness of the torsion free rank is sufficient for spectral synthesis on an Abelian group? At this moment we have that spectral synthesis is a strictly stronger property of the group than spectral analysis: by the results of the previous chapter it follows that there are Abelian groups on which spectral analysis holds and spectral synthesis fails to hold. For instance, the free Abelian group of rank ω is one of them. Another natural question is the following: are there any Abelian groups of infinite rank possessing spectral synthesis? We shall answer these questions in the forthcoming paragraphs.

15.3 Spectral synthesis on Abelian groups

Obviously, spectral synthesis holds on finite Abelian groups. By the result 15.4, and by Theorem 15.8 we have the following result.

Theorem 15.14. *Spectral synthesis holds on every finitely generated Abelian group.*

Proof. Indeed, every finitely generated Abelian group is the homomorphic image of a free Abelian group of finite rank (see e.g. [Fuchs (1970)], Corollary 14.3, p. 74), which is isomorphic to $\mathbb{Z}^n$ for some natural number n. Hence our statement is a consequence of Theorem 15.4 and of Theorem 15.8. □

The following result is related to Theorem 14.4 (see [Bereczky and Székelyhidi (2005)]).

Theorem 15.15. *Spectral synthesis holds on every Abelian torsion group.*

Proof. We show that if G is a torsion group, then every ideal in $\mathbb{C}G$ is the intersection of maximal ideals. First of all, it is a simple observation that for such a group, every prime ideal of $\mathbb{C}G$ is a maximal ideal. For if P is a prime ideal in $\mathbb{C}G$ and g is an element of G, then

$$0 = g^k - 1 = (g-1)(g-\eta)\cdots(g-\eta^{k-1})$$

belongs to P with some positive integer k and a complex primitive k-th root of unity η, hence some $g - \eta^m$ is in P. In particular, g is in $P + \mathbb{C}$, and this is true for all the elements of G, so $\mathbb{C}G/P = (P+\mathbb{C})/P \cong \mathbb{C}$. However, having a field as quotient ring, P has to be maximal. This argument also shows that every maximal ideal in $\mathbb{C}G$ is exponential, however, this follows from Theorem 14.4.

Now since all prime ideals are maximal, our statement reduces to the property that every ideal in the group algebra $\mathbb{C}G$ is an intersection of prime ideals, that is, every ideal is *semi-prime*. However, this property is a known characterization of the so-called fully idempotent rings, which are, by definition, the rings, where $I^2 = I$ for all ideals I. (For more details and a proof of this characterization, see e.g. [Wisbauer (1991)].) In particular, a regular ring is always fully idempotent (as if x is in the ideal I of the ring R, and it satisfies $x = xyx$ with some ring element y, then clearly x belongs to I^2), and by a theorem of M. Auslander (see [Auslander (1957)]), a commutative complex group algebra $\mathbb{C}G$ is regular if and only if G is a torsion group.

Hence, if G is a torsion group, then every ideal in $\mathbb{C}G$ is the intersection of exponential maximal ideals, which implies that each variety on G is the closure of the set of all exponentials in the variety. Now our proof is complete. □

This theorem shows that there are non-finitely generated Abelian groups possessing spectral synthesis. The following theorem completely characterizes discrete Abelian groups with spectral synthesis.

Theorem 15.16. *(M. Laczkovich–L. Székelyhidi) Spectral synthesis holds on an Abelian group if and only if its torsion free rank is finite.*

The proof of this theorem depends on a series of lemmas. The reader is referred to [Laczkovich and Székelyhidi (2007)]. We note that this theorem obviously implies the following result.

Theorem 15.17. *If spectral synthesis holds on two Abelian groups, then it holds on their direct sum, too.*

Conversely, this theorem, together with the theorem of Lefranc 15.4 and Theorem 15.15, implies Theorem 15.16, as it is easy to see. Unfortunately, a simple direct proof of Theorem 15.17 has not been found so far.

Bibliography

Abramovich, Y. A., Aliprantis, C. D., Sirotkin, G. and Troitsky, V. G. (2005). Some open problems and conjectures associated with the invariant subspace problem, *Positivity* **9**, 3, pp. 273–286.

Aczél, J. (1966). *Lectures on functional equations and their applications*, Mathematics in Science and Engineering, Vol. 19 (Academic Press, New York).

Anselone, P. M. and Korevaar, J. (1964). Translation invariant subspaces of finite dimension, *Proc. Amer. Math. Soc.* **15**, pp. 747–752.

Arveson, W. (2002). *A short course on spectral theory*, *Graduate Texts in Mathematics*, Vol. 209 (Springer-Verlag, New York).

Atiyah, M. F. and Macdonald, I. G. (1969). *Introduction to commutative algebra* (Addison-Wesley Publishing Co., Reading, Mass.-London-Don Mills, Ont.).

Auslander, M. (1957). On regular group rings, *Proc. Amer. Math. Soc.* **8**, pp. 658–664.

Benedetto, J. J. (1975). *Spectral synthesis* (Academic Press Inc. [Harcourt Brace Jovanovich Publishers], New York).

Bereczky, Á. and Székelyhidi, L. (2005). Spectral synthesis on torsion groups, *J. Math. Anal. Appl.* **304**, 2, pp. 607–613.

Bhatt, S. J. and Dedania, H. V. (2005). A note on generalized characters, *Proc. Indian Acad. Sci. Math. Sci.* **115**, 4, pp. 437–444.

Bourbaki, N. (1972). *Elements of mathematics. Commutative algebra* (Hermann, Paris).

Bourbaki, N. (1998a). *General topology. Chapters 1–4*, Elements of Mathematics (Berlin) (Springer-Verlag, Berlin).

Bourbaki, N. (1998b). *General topology. Chapters 5–10*, Elements of Mathematics (Berlin) (Springer-Verlag, Berlin).

Chandrasekharan, K. (2011). *A course on topological groups*, *Texts and Readings in Mathematics*, Vol. 9 (Hindustan Book Agency, New Delhi).

Dixmier, J. (1969). *Les C^*-algèbres et leurs représentations*, Deuxième édition. Cahiers Scientifiques, Fasc. XXIX (Gauthier-Villars Éditeur, Paris).

Djokovič, D. Ž. (1969/1970). A representation theorem for $(X_1 - 1)(X_2 - 1)\cdots(X_n - 1)$ and its applications, *Ann. Polon. Math.* **22**, pp. 189–198.

Dummit, D. S. and Foote, R. M. (2004). *Abstract algebra*, 3rd edn. (John Wiley

& Sons Inc., Hoboken, NJ).

Dunford, N. and Schwartz, J. T. (1988a). *Linear operators. Part I*, Wiley Classics Library (John Wiley & Sons Inc., New York).

Dunford, N. and Schwartz, J. T. (1988b). *Linear operators. Part II*, Wiley Classics Library (John Wiley & Sons Inc., New York).

Dunford, N. and Schwartz, J. T. (1988c). *Linear operators. Part III*, Wiley Classics Library (John Wiley & Sons Inc., New York).

Edwards, R. E. (1979). *Fourier series. A modern introduction. Vol. 1*, *Graduate Texts in Mathematics*, Vol. 64, 2nd edn. (Springer-Verlag, New York).

Edwards, R. E. (1982). *Fourier series. Vol. 2*, *Graduate Texts in Mathematics*, Vol. 85, 2nd edn. (Springer-Verlag, New York).

Ehrenpreis, L. (1955). Mean periodic functions. I. Varieties whose annihilator ideals are principal, *Amer. J. Math.* **77**, pp. 293–328.

Elliott, R. J. (1965). Two notes on spectral synthesis for discrete Abelian groups, *Proc. Cambridge Philos. Soc.* **61**, pp. 617–620.

Engert, M. (1970). Finite dimensional translation invariant subspaces, *Pacific J. Math.* **32**, pp. 333–343.

Fréchet, M. (1909). Une définition fonctionelle des polynômes, *Nouv. Ann.* **49**, pp. 145–162.

Fuchs, L. (1958). *Abelian groups* (Publishing House of the Hungarian Academy of Sciences, Budapest).

Fuchs, L. (1970). *Infinite abelian groups. Vol. I*, Pure and Applied Mathematics, Vol. 36 (Academic Press, New York).

Gajda, Z. (1984). Additive and convex functions in linear topological spaces, *Aequationes Math.* **27**, 3, pp. 214–219.

Gajda, Z. (1987). A solution to a problem of J. Schwaiger, *Aequationes Math.* **32**, 1, pp. 38–44.

Greenleaf, F. P. (1969). *Invariant means on topological groups and their applications*, Van Nostrand Mathematical Studies, No. 16 (Van Nostrand Reinhold Co., New York).

Grothendieck, A. (1973). *Topological vector spaces* (Gordon and Breach Science Publishers, New York).

Gurevič, D. I. (1975). Counterexamples to a problem of L. Schwartz, *Funkcional. Anal. i Priložen.* **9**, 2, pp. 29–35.

Halmos, P. R. (1950). *Measure Theory* (D. Van Nostrand Company, Inc., New York, N. Y.).

Halmos, P. R. (1974a). *Finite-dimensional vector spaces*, 2nd edn. (Springer-Verlag, New York).

Halmos, P. R. (1974b). *Naive set theory* (Springer-Verlag, New York).

Halmos, P. R. (1995). *Linear algebra problem book*, *The Dolciani Mathematical Expositions*, Vol. 16 (Mathematical Association of America, Washington, DC).

Hausdorff, F. (1962). *Set theory*, Second edition. Translated from the German by John R. Aumann et al (Chelsea Publishing Co., New York).

Hewitt, E. and Ross, K. A. (1970). *Abstract harmonic analysis. Vol. II: Structure and analysis for compact groups. Analysis on locally compact Abelian groups*, Die Grundlehren der mathematischen Wissenschaften, Band 152 (Springer-Verlag, New York).

Hewitt, E. and Ross, K. A. (1979). *Abstract harmonic analysis. Vol. I, Grundlehren der Mathematischen Wissenschaften [Fundamental Principles of Mathematical Sciences]*, Vol. 115, 2nd edn. (Springer-Verlag, Berlin).

Holmes, R. B. (1975). *Geometric functional analysis and its applications* (Springer-Verlag, New York).

Isaacs, I. M. (2006). *Character theory of finite groups* (AMS Chelsea Publishing, Providence, RI).

Izzo, A. J. (1992). A functional analysis proof of the existence of Haar measure on locally compact abelian groups, *Proc. Amer. Math. Soc.* **115**, 2, pp. 581–583.

Jacobson, N. (1953). *Lectures in Abstract Algebra. Vol. II. Linear algebra* (D. Van Nostrand Co., Inc., Toronto-New York-London).

Jacobson, N. (1985). *Basic algebra. I*, 2nd edn. (W. H. Freeman and Company, New York).

Jacobson, N. (1989). *Basic algebra. II*, 2nd edn. (W. H. Freeman and Company, New York).

Jarchow, H. (1981). *Locally convex spaces* (B. G. Teubner, Stuttgart).

Jech, T. J. (1973). *The axiom of choice* (North-Holland Publishing Co., Amsterdam).

Kelley, J. L. (1975). *General topology* (Springer-Verlag, New York).

Kelley, J. L. and Namioka, I. (1976). *Linear topological spaces* (Springer-Verlag, New York).

Köthe, G. (1969). *Topological vector spaces. I*, Translated from the German by D. J. H. Garling. Die Grundlehren der mathematischen Wissenschaften, Band 159 (Springer-Verlag New York Inc., New York).

Köthe, G. (1979). *Topological vector spaces. II, Grundlehren der Mathematischen Wissenschaften [Fundamental Principles of Mathematical Science]*, Vol. 237 (Springer-Verlag, New York).

Kuczma, M. (2009). *An introduction to the theory of functional equations and inequalities*, 2nd edn. (Birkhäuser Verlag, Basel).

Kuratowski, K. (1962). *Introduction to set theory and topology*, Translated from the revised Polish edition by Leo F. Boron (Pergamon Press, Ltd., Oxford).

Laczkovich, M. (2000/01). Exponential polynomials, *Mat. Lapok (N.S.)* **10**, 1, pp. 22–31 (2004).

Laczkovich, M. (2004). Polynomial mappings on abelian groups, *Aequationes Math.* **68**, 3, pp. 177–199.

Laczkovich, M. and Székelyhidi, G. (2005). Harmonic analysis on discrete abelian groups, *Proc. Amer. Math. Soc.* **133**, 6, pp. 1581–1586.

Laczkovich, M. and Székelyhidi, L. (2007). Spectral synthesis on discrete abelian groups, *Math. Proc. Cambridge Philos. Soc.* **143**, 1, pp. 103–120.

Laird, P. G. (1979). On characterizations of exponential polynomials, *Pacific J. Math.* **80**, 2, pp. 503–507.

Laird, P. G. and McCann, R. (1984). On some characterizations of polynomials, *Amer. Math. Monthly* **91**, 2, pp. 114–116.

Larsen, R. (1973). *Functional analysis: an introduction* (Marcel Dekker Inc., New York).

Lefranc, M. (1958). Analyse spectrale sur Z_n, *C. R. Acad. Sci. Paris* **246**, pp. 1951–1953.

Loomis, L. H. (1953). *An introduction to abstract harmonic analysis* (D. Van Nostrand Company, Inc., Toronto-New York-London).

Luong, B. (2009). *Fourier analysis on finite abelian groups*, Applied and Numerical Harmonic Analysis (Birkhäuser Boston Inc., Boston, MA).

Maak, W. (1967). *Fastperiodische Funktionen*, Zweite, korrigierte Auflage. Die Grundlehren der mathematischen Wissenschaften, Band 61 (Springer-Verlag, Berlin).

Malgrange, B. (1954). Sur quelques propriétés des équations de convolution, *C. R. Acad. Sci. Paris* **238**, pp. 2219–2221.

Matsumura, H. (1980). *Commutative algebra, Mathematics Lecture Note Series*, Vol. 56, 2nd edn. (Benjamin/Cummings Publishing Co., Inc., Reading, Mass.).

Matsumura, H. (1986). *Commutative ring theory, Cambridge Studies in Advanced Mathematics*, Vol. 8 (Cambridge University Press, Cambridge).

Mazur, S. and Orlicz, W. (1934a). Grundlegende Eigenschaften der polynomischen Operationen I.,, *Studia Math.* **5**, pp. 50–68.

Mazur, S. and Orlicz, W. (1934b). Grundlegende Eigenschaften der polynomischen Operationen II.,, *Studia Math.* **5**, pp. 179–189.

McKiernan, M. A. (1967). On vanishing nth ordered differences and Hamel bases, *Ann. Polon. Math.* **19**, pp. 331–336.

McKiernan, M. A. (1977a). Equations of the form $H(x \circ y) = \sum_i f_i(x)g_i(y)$, *Aequationes Math.* **16**, 1-2, pp. 51–58.

McKiernan, M. A. (1977b). The matrix equation $a(x \circ y) = a(x) + a(x)a(y) + a(y)$, *Aequationes Math.* **15**, 2-3, pp. 213–223.

Mendelson, B. (1990). *Introduction to topology*, 3rd edn., Dover Books on Advanced Mathematics (Dover Publications Inc., New York).

Montgomery, D. and Zippin, L. (1974). *Topological transformation groups* (Robert E. Krieger Publishing Co., Huntington, N.Y.).

Moore, G. H. (1982). *Zermelo's axiom of choice, Studies in the History of Mathematics and Physical Sciences*, Vol. 8 (Springer-Verlag, New York).

Munkres, J. R. (1975). *Topology: a first course* (Prentice-Hall Inc., Englewood Cliffs, N.J.).

Nagata, M. (1975). *Local rings* (Robert E. Krieger Publishing Co., Huntington, N.Y.).

Newman, M. (1967). Two classical theorems on commuting matrices, *Jour. of Res. Math. and Math. Phys.* **71B**, 2–3, pp. 69–71.

Pontryagin, L. S. (1966). *Topological groups*, Translated from the second Russian edition by Arlen Brown (Gordon and Breach Science Publishers, Inc., New York).

Reiter, H. (1968). *Classical harmonic analysis and locally compact groups* (Clarendon Press, Oxford).

Rudin, W. (1964). *Principles of mathematical analysis*, Second edition (McGraw-

Hill Book Co., New York).
Rudin, W. (1987). *Real and complex analysis*, 3rd edn. (McGraw-Hill Book Co., New York).
Rudin, W. (1990). *Fourier analysis on groups*, Wiley Classics Library (John Wiley & Sons Inc., New York).
Rudin, W. (1991). *Functional analysis*, 2nd edn., International Series in Pure and Applied Mathematics (McGraw-Hill Inc., New York).
Schwartz, L. (1947). Théorie générale des fonctions moyenne-périodiques, *Ann. of Math. (2)* **48**, pp. 857–929.
Shulman, E. (2011). Some extensions of the Levi-Civitá functional equation and richly periodic spaces of functions, *Aequationes Math.* **81**, 1-2, pp. 109–120.
Spindler, K. (1994a). *Abstract algebra with applications. Vol. I* (Marcel Dekker Inc., New York).
Spindler, K. (1994b). *Abstract algebra with applications. Vol. II* (Marcel Dekker Inc., New York).
Stone, J. J. (1960). Exponential polynomials on commutative semigroups, *Appl. Math. and Stat. Lab. Technical Note, Stanford University* **14**.
Stroppel, M. (2006). *Locally compact groups*, EMS Textbooks in Mathematics (European Mathematical Society (EMS), Zürich).
Székelyhidi, L. (1979). Remark on a paper of M. A. McKiernan: "On vanishing nth-ordered differences and Hamel bases" [Ann. Polon. Math. **19** (1967), 331–336; MR **36** #4183], *Ann. Polon. Math.* **36**, 3, pp. 245–247.
Székelyhidi, L. (1982a). Note on exponential polynomials, *Pacific J. Math.* **103**, 2, pp. 583–587.
Székelyhidi, L. (1982b). On a class of linear functional equations, *Publ. Math. Debrecen* **29**, 1-2, pp. 19–28.
Székelyhidi, L. (1988). Fréchet's equation and Hyers theorem on noncommutative semigroups, *Ann. Polon. Math.* **48**, 2, pp. 183–189.
Székelyhidi, L. (1991). *Convolution type functional equations on topological abelian groups* (World Scientific Publishing Co. Inc., Teaneck, NJ).
Székelyhidi, L. (1999). On convolution type functional equations, *Math. Pannon.* **10**, 2, pp. 271–275.
Székelyhidi, L. (2000). On the extension of exponential polynomials, *Math. Bohemica* **125**, 3, pp. 365–370.
Székelyhidi, L. (2001). A Wiener Tauberian theorem on discrete abelian torsion groups, *Annales Acad. Paedag. Cracov., Studia Mathematica I.* **4**, pp. 147–150.
Székelyhidi, L. (2004). The failure of spectral synthesis on some types of discrete abelian groups, *J. Math. Anal. Appl.* **291**, 2, pp. 757–763.
Székelyhidi, L. (2005). Polynomial functions and spectral synthesis, *Aequationes Math.* **70**, 1-2, pp. 122–130.
Székelyhidi, L. (2012a). Noetherian rings of polynomial functions on Abelian groups, *Aequationes Math.* **84**, 1-2, pp. 41–50.
Székelyhidi, L. (2012b). Polynomial functions and spectral synthesis on abelian groups, *Banach J. Math. Anal.* **6**, 1, pp. 124–131.
Terras, A. (1999). *Fourier analysis on finite groups and applications, London*

Mathematical Society Student Texts, Vol. 43 (Cambridge University Press, Cambridge).

van der Lijn, G. (1940a). Les polynomes abstraits. I, *Bull. Sci. Math.* **64**, pp. 55–80.

van der Lijn, G. (1940b). Les polynomes abstraits (Suite II), *Bull. Sci. Math. (2)* **64**, pp. 128–144.

van der Waerden, B. L. (1991a). *Algebra. Vol. I* (Springer-Verlag, New York).

van der Waerden, B. L. (1991b). *Algebra. Vol. II* (Springer-Verlag, New York).

Wisbauer, R. (1991). *Foundations of module and ring theory*, *Algebra, Logic and Applications*, Vol. 3, german edn. (Gordon and Breach Science Publishers, Philadelphia, PA).

Yadav, B. S. (2005). The invariant subspace problem, *Nieuw Arch. Wiskd. (5)* **6**, 2, pp. 148–152.

Yosida, K. (1995). *Functional analysis*, Classics in Mathematics (Springer-Verlag, Berlin).

Zariski, O. and Samuel, P. (1975a). *Commutative algebra. Vol. I* (Springer-Verlag, New York).

Zariski, O. and Samuel, P. (1975b). *Commutative algebra. Vol. II* (Springer-Verlag, New York).

Zorn, M. (1935). A remark on method in transfinite algebra, *Bull. Amer. Math. Soc.* **41**, 10, pp. 667–670.

Index

Printed in the United States
By Bookmasters